ON PURPOSE

On Purpose

Michael Ruse

PRINCETON UNIVERSITY PRESS
PRINCETON & OXFORD

Published by Princeton University Press,
41 William Street, Princeton, New Jersey 08540

In the United Kingdom: Princeton University Press,
6 Oxford Street, Woodstock, Oxfordshire OX20 1TR

press.princeton.edu

Jacket design by Kimberly Glyder

ISBN 978-0-691-17246-0

British Library Cataloging-in-Publication Data is available

This book has been composed in Miller

Printed on acid-free paper. ∞

Printed in the United States of America

10 9 8 7 6 5 4 3 2 1

For Joe and Celine

CONTENTS

ACKNOWLEDGMENTS

THIS PROJECT started nearly half a century ago when I was writing my first book, *The Philosophy of Biology* (1973). Framed very much in the school of "logical empiricism"—the leaders were two men whose names and memory I still revere, Ernest Nagel and Carl "Peter" Hempel—everything was going along swimmingly until I got to the chapter on function or purpose. Something went wrong, for I could not fit the discussion into the mold, especially the mold of science as a value-free inquiry, as an enterprise that drains itself of the human element—in Karl Popper's felicitous phrase, "knowledge without a knower." Eventually, I plowed on, or rather through, and the book was finished and published. But the problem of purpose kept nagging away—even back then I think I had insights into the way things had to go—and it has been a lifetime's quest for understanding, frustrating at times but incredibly invigorating. Now I think I know the answer, and it is here in this book, the summing up of a fifty-year obsession with the problem.

So, first, I want to thank Nagel and Hempel for setting me off on this quest. It was from them, as well as from others in the field, I learned that philosophy never stands still; there is always work to be done, criticizing and extending. An insight that never withers is that you learn most from those with whom you disagree most, and I very much hope that this book exemplifies this truth. I am very grateful to my fellow philosophers who have so stimulated me. In a rather different way, I want to thank Plato and Aristotle, Kant and Darwin. As you will see, my quest has taken me back to their writings. It has been a great privilege to spend time with minds such as these. If my huge respect for and sheer enjoyment and excitement at what they produced does not come across on every page, then I have failed myself, I have failed you the reader, and, most sadly, I have failed them. I want this book read in a positive manner. I shall have critical things to say but always in the sense of wanting to move the conversation forward.

At the immediate level, my thanks above all go to my editor at Princeton University Press, Rob Tempio. When he first asked me to write this book, I agreed, believing that it would be a good way of summing up ideas about which I have been thinking and writing for many years. Deftly, he steered me toward imposing on my material a new and, I think, informative framework, looking at old problems in a way that hitherto I had grasped but vaguely. At times, responding to Rob's comments, as well as to those of the referees he so astutely chose, I wondered why he and they didn't simply write the book themselves. Socrates once said he was the wisest man alive since he knew one thing, namely, that he knew nothing. Yes, indeed.

I am most grateful to my colleagues, Nat Stein and Randy Clarke, for looking at an earlier version of my manuscript and giving me useful comments. My dear friend, professor at Fordham University and the English-speaking world's foremost authority on Thomas Aquinas, Brian Davies, pointed to the egregious mistakes I made about the great Christian philosophers. Given the melancholic pleasure that this gave him, I am sure that there is lurking there a new ontological argument about my necessary existence. I am deeply in the debt of my student, assistant, and, above all, friend Jeff O'Connor for doing more of the humdrum jobs than one has any right to ask of any person. As always, I pay grateful memory to William and Lucyle Werkmeister, whose legacy made possible my professorship and the research funds attached to it. And finally, above all, I celebrate the love and warm companionship of my family, especially my daughter, Emily, whom you will be meeting, and my wife, Lizzie. You will later learn the context when I say that the trade-off for spending so much time being antisocial and writing this book is that Nutmeg arrived into our lives.

PROLOGUE

Our little Kinsmen—after Rain
In plenty may be seen,
A Pink and Pulpy multitude
The tepid Ground upon.

A needless life, it seemed to me
Until a little Bird
As to a Hospitality
Advanced and breakfasted -

As I of He, so God of Me
I pondered, may have judged,
And left the little Angle Worm
With Modesties enlarged.

—EMILY DICKINSON,
WRITTEN ABOUT 1864

"A NEEDLESS LIFE." A life without worth. A life without purpose, until the poet saw that the worm was breakfast for the little bird. This notion of "purpose"—understanding or doing something in the light of the ends that it serves—is interesting and complex. We don't generally use this kind of thinking—function talk, making reference to what Aristotelians called "final causes" and what, since the Enlightenment, has been dubbed "teleological" thinking—in the physical world, the world of planets and pendulums, of protons and plate tectonics. No one would ask about the purpose of the meteorite that smashed into the earth some sixty-six million years ago, creating such a hostile atmosphere that that was the end of the dinosaurs. It just happened. There was no purpose to it.

In this meteorite case, obviously we are talking about organisms—the unfortunate dinosaurs. We are not talking about the organisms in their own right but rather as what happened to them as the consequence of certain physical events. The impact first brought on a huge rise in the atmosphere's carbon dioxide levels and massive heating, and then this was followed by a kind of nuclear winter as the dust in the air blocked the sunlight and caused extreme cold. The knock-on effect was to destroy the earth's vegetation, and hence the wretched brutes starved to death.[1] But when we turn to the world of organisms in its own right—wanting to understand how living beings work—we ask about purpose all of the time, as when, for example (to stay with dinosaurs), we ask a question like: What is the purpose of those funny, pointed, finlike appendages ("plates") all down the back of the stegosaurus? We see such use also in the human world as when, for example, we ask a student: Why are you taking a course in calculus rather than in Elizabethan poetry? In the meteor case, at a causal level, we are interested in and only in the prior causes. How did the impact bring on a rise in the carbon dioxide level and so forth? We are not asking about how the dinosaur deaths brought on the impact. In the cases of organisms and humans, we are interested not so much in the prior causes behind the things or actions—presumably in the stegosaurus case these involve certain physiological processes and in the student case probably filling in one set of spaces in a computerized questionnaire rather than others—but in the hoped-for results, the goals. For the stegosaurus, it is thought that the purpose is temperature control—radiating heat when the brute is too hot and soaking up the sun's rays when the brute is too cool.[2] For the student, the goal is to get into medical school and lead a fulfilling life rather than end up as an adjunct humanities professor in some part of the world hitherto unknown to civilization.

What makes the whole situation complex and interesting is that, in some real sense, purpose-questions make reference to the future. Note that it is not just a matter of the future being involved. In the physical sciences we think a lot about the future. I am sure that there are today many earth scientists, hunched over their computers, building models about the effects of impacts. But it is prior causes alone—what are often known as "efficient causes"—to which appeal is made. As one might say, it is turtles all of the way down. Which physical events brought on which physical events and in what fashion, and what are the likelihoods of it all happening again? Likewise, in the case of normal causal processes like physiological development or computer form filling, it is all a matter of efficient causes. But here's the rub! If the dinosaurs died, if the stegosaurus has the weird appendages, if the student has the admission form, you know that the factors bringing all of this about must have occurred—either in the past or at most at the same time as the event or object being studied. In the purpose case, the intended future may never occur. The stegosaurus might get swept away by a flood before it can use its plates; the student might change his or her mind and become something socially valuable, like a stockbroker or banker. What then is going on and why do we keep using purpose-type understanding? Why do we still seek final causes? How can the possibly nonexistent be a determining factor?

Let us, you say, get away from all of this. Let us stick with efficient causes. After all, that is what happens in the physical sciences. In the 1970s, the English inventor James Lovelock and the American cell biologist Lynn Margulis—both in their own rights very distinguished scientists—proposed the Gaia hypothesis, the claim that Planet Earth is an organism and because of this maintains homeostasis in the face of external changes and disruptions.[3] From the heart of their own community, loud was the cry of "pseudoscience," in main part because the hypothesis was (with good reason) judged teleological.[4] The Gaia enthusiasts argued that the earth behaves as it does—absorbing salt from the oceans, for instance—in order to remain stable and capable of sustaining living beings. Such thinking for the keepers of the sacred flame of proper thinking—the Richard Dawkinses of the world—was nigh heretical, to be banished to the outer circles of the universe, the phantom zone, out of sight and forgotten forever.[5] Had not the great Francis Bacon likened final causes to Vestal Virgins, decorative but sterile?

You naive fool, comes the response. Physicists and wannabes may strut and fret their hour upon the stage, but then they are heard no more. As the philosopher Immanuel Kant argued forcibly in his *Critique of the Power of Judgment*,[6] in the world of organisms, not to mention that of humans, you may try and try to get away from purpose-filled talk, but you are never, ever going to succeed. Like taxes and death, it is one of those things that humans are stuck with, a lover that you fear to parade in public and yet you cannot live without. And not only is the reason for that the spur to an interesting question, but it opens a cascade of other interesting questions. You can cheat on your taxes. Ultimately, for all that Antonius Block did his best in the *Seventh Seal*, you can never cheat on death. Which raises the biggest purpose-laden questions of them all: Why are we here? Where are we going? What is the purpose of it all? Can I now make any difference to the future? Or are we like the little worm in Emily Dickinson's chilly poem? We may have a purpose in the cosmic scheme of things, but it is not necessarily one directed to our personal well-being or happiness.

I am an evolutionist. I do not think that this necessarily implies that the present is better than the past—this, indeed, is one of the questions we shall be asking. I do think that the present can be understood only by knowledge of the past. Hence, my approach in this short monograph is that of the historian of ideas. I want to dig back to the origins of purpose thinking and then move through history to the present. As you can well imagine, there is a huge literature on the people I discuss. As in physics, to every action there is an equal and opposite reaction, to every claim about Plato and Aristotle and the others, there is an equal and opposite counterclaim. That is how we scholars make our livings. I have just cowritten a book with a good friend where we offer completely different interpretations of the work of Charles Darwin, me seeing him entirely in the British tradition and my cowriter seeing him as a German Romantic in everything but his language of birth.[7] Fortunately, in the tradition of my great predecessors, notably Arthur Lovejoy and Isaiah Berlin, I tell a tale here not for the sake of history itself but for the sake of philosophy. For the sake of a problem today. Thus, I can be cavalier with the voluminous material, picking out (not idiosyncratically) claims and judgments pertinent to my end. Why do we use purpose talk? Could we eliminate it? Should we eliminate it? Is it a burdensome relic of the past, like the appendix, as it was long taken to be, or does it have yet a vital place in our thinking, like the brain itself? As always in the Western context when asking such questions as these, we are directed back to two sources: Athens and Jerusalem. In our terms, to the start of philosophy and to the start of religion. The early Christian writer Tertullian (155–240), who made this distinction, warned against Athens. This, as we shall see, is one of the things with which we shall wrestle. For now, take the two sources and, with this as our guide, let us begin our tale.

ON PURPOSE

CHAPTER ONE

Athens

ARISTOTLE (384–322 BC) SAID, “Knowledge is the object of our inquiry, and men do not think they know a thing till they have grasped the ‘why’ of it (which is to grasp its primary cause).”[1] He was not the first to raise the question of causation, for it was nigh an obsession of his philosophical predecessors, back through his teacher Plato (ca. 429–347 BC) to Socrates (469–399 BC), and to the earlier “pre-Socratic” thinkers, including Empedocles (ca. 495–435 BC), Anaxagoras (ca. 510–ca. 428 BC), and the atomist Democritus (ca. 460–ca. 370 BC). They all grasped that in some sense causation—what it is that makes things happen—is (or is often taken to be) both a backward-looking matter and a forward-looking matter.[2] The nail is driven into the piece of wood. Backward-looking in the sense that this happens because a hammer was picked up and used to hit the head of the nail; forward-looking in the sense that this happens because the builder wants to tie the planks together to support a roof. The builder did this “on purpose” or “for a purpose.” He wanted that end. A roof was something of *value* to him.

I shall argue that this forward-looking side to causation—the subject of our inquiry—lends itself to three different approaches. I do not pretend to originality in spotting these approaches. Others, for instance, R. J. (Jim) Hankinson and Thomas Nagel, have certainly remarked on this triune side.[3] It is in tracing the way that it persists that makes the story so interesting and illuminating. The first approach, often known as "external" teleology, is the most obvious and intuitively plausible. It involves a mind, whether human or divine or something else. "For God so loved the world, that he gave his only begotten Son, that whosoever believeth in him should not perish, but have everlasting life" (John 3:16).[4] God, right now, let Jesus die on the cross, so that you, the sinner, should have everlasting life in the future. The second approach, often known as "internal teleology," is a bit trickier. It involves a kind of life force in some sense, something that need not be conscious, and actually in the broader sense need not even be alive. It might be more a kind of principle of ordering about the world, something that makes everything essentially end-directed. When we see it being argued for, we shall get a better sense of what it is all about. These two notions of purpose, of teleology, go back readily to the Greeks. The third kind of approach we might call "eliminative" or, more positively, "heuristic" teleology, seeing forward-looking causation—purpose—as in some sense purely conceptual, something we might use to understand the world but in no sense constitutive of the world. This label would apply to—or at least is anticipated in—the approach of Democritus and comes out more vividly in the (several centuries later) poetry of the Roman Lucretius (ca. 99–ca. 55 BC). But it is not until the modern era that this approach could be developed fully.

With respect to the first two approaches, it is not always easy to tell if one has either external or internal teleology. In Emily Dickinson's poem, is there a designer god behind everything or is it all a matter of an impersonal force, an Immanent Will (as it has sometimes been called)? What we can say is that Plato offered the first full discussion of external teleology and Aristotle the first full discussion of internal teleology, with the atomists at the least the forerunners of the heuristic option.

Plato

There are two main sources for Plato's thinking about purpose, about teleology. The first is in the *Phaedo*, the dialogue about Socrates's last day on Earth. It is a middle dialogue, and given the nature of the discussion is generally considered a vehicle for Plato's own thinking—apart from anything else, Plato notes explicitly that he himself was not present, which gives us a clue that there has to be some element of creativity—although there is a comparable discussion attributed to Socrates himself by Xenophon (ca. 430–354 BC), and a version of the argumentation may go back to Anaxagoras. Surrounded by the young men who are his followers, much of the discussion Socrates directs is (hardly surprisingly) about key issues, such as the nature of the soul—more on this shortly—and questions about existence beyond this life. Almost in passing, Socrates raises the question of the deity. It is not so much a question of offering a formal proof but in showing how we need such a concept in order to make sense of the ways in which we understand things.

Normally, such an issue doesn't arise. "I thought before that it was obvious to anybody that men grew through eating and drinking, for food adds flesh to flesh and bones to bones, and in the same way appropriate parts were added to all other parts of the body, so that the man grew from an earlier small bulk to a large bulk later, and so a small man became big."[5] This is backward-looking causation, that is, what we have seen called "efficient causation." Plato acknowledges that this is not a bad explanation—we do get bigger thanks to eating and drinking—but it is in some sense incomplete. Why would one bother to eat and drink? Why would one want to grow and put on weight? See here how the notion of *value* is coming in. What is the point of doing something? What's the purpose? Why do we want the end result? Here we need to switch to forward-looking causation or (what within the Aristotelian system was called) "final cause." We are—or rather will be—better off if we grow. This is crucial.

Something happens that we *value*. Which is just fine and dandy, but why should it happen? Why doesn't eating and drinking make us lose weight? "One day I heard someone reading, as he said, from a book of Anaxagoras, and saying that it is Mind that directs and is the cause of everything. I was delighted with this cause and it seemed to me to be good, in a way, that Mind should be the cause of all. I thought that if this were so, the directing Mind would direct everything and arrange each thing in the way that was best."[6] So now one has a guide to discovery. "Then if one wished to know the cause of each thing, why it comes to be or perishes or exists, one had to find what the best way was for it to be, or to be acted upon, or to act."[7]

Note that we have a heuristic here but more than this, although it is more a presupposition than an explicit proof. Things don't just happen. They are ordered for the best, and this is done by a mind—or rather by a Mind. The teleology in this sense is external—imposed upon the world from without.

Atomist Interlude

Pause for a moment, to dig a little more deeply. You have features, let us say teeth or hands or whatever. These are brought about through efficient causes, the physiological effects of eating and drinking. They also have purposes or ends, what we are going to call final causes. These features are of value. And God or a Mind is being invoked to explain everything. The Design Argument, although note that truly we have a two-part argument here. First, *to* the design-like nature of the world. Second, *from* this nature to a God. Plato more or less takes the first part of the argument as a given and is focused on the second part. Aristotle, as we shall see, as a sometime very serious biologist, has more focus on the first part. In order to bring out these two moves, let me make continued use of the fact that I am not now writing a straight history of philosophy but a history of ideas directed toward the present, and so I have greater freedom to move back and forth in time. Interrupt Plato and turn for a moment for contrast and illumination to the atomists. They argued that we have an infinite universe, infinite time, and nothing but particles—atoms—buzzing around in space or the void. Every now and then they will join up, and first we have disembodied parts—an eye here and a leg there. In the words of Lucretius, writing in the tradition of the materialist Epicurus (341–270 BC), who in his physics followed Democritus:

> At that time the earth tried to create many monsters
> with weird appearance and anatomy—
> androgynous, of neither one sex nor the other but
>   somewhere in between; some footless, or handless;
> many even without mouths, or without eyes and blind;
> some with their limbs stuck together all along their body,
> and thus disabled from doing harm or obtaining anything
>   they needed.
> These and other monsters the earth created.
> But to no avail, since nature prohibited their development.
> They were unable to reach the goal of their maturity,
> to find sustenance or to copulate.[8]

Thus far, nothing works. It is just a mess, of no value whatsoever. But then, given infinite time, things joined up in functioning ways.

> First, the fierce and savage lion species
> has been protected by its courage, foxes by cunning, deer by
>   speed of flight. But as for the light-sleeping minds of
>   dogs, with their faithful heart,
> and every kind born of the seed of beasts of burden,
> and along with them the wool-bearing flocks and the
>   horned tribes,
> they have all been entrusted to the care of the human race
>   (5.862–67)

This obviously is a direct challenge to the second move in the Design argument. The immediate objects of Lucretius's poem are probably the Stoics (see below), but Plato is in direct line. In the *Sophist*, having invoked a deity to explain the design-like nature of everything, animals, plants, the earth itself, he asks bluntly, "Are we going to say that nature produces them by some spontaneous cause that generates them without any thought, or by a cause that works by reason and divine knowledge derived from a god?"[9] The first disjunct is the atomist's happy reply. No need to invoke a god or whatever to explain purpose—"intelligence, along with color, flavor, and innumerable other attributes, is among the properties that supervene on complex structures of atoms and the void."[10] Which in turn rather implies that the atomists allow the first part of the argument. Things, organisms in particular, show purpose. They have features serving their ends (fierceness, cunning, running ability) or our ends (faithfulness, strength for work, wool coats, milk and meat). Lucretius admits this, but reluctantly. It is certainly not part of reality. Eyes were not made for seeing or legs for keeping us upright. It is rather that the eyes and legs appeared and then were put to use. To think otherwise is to get things backward.

> All other explanations of this type which they offer
> are back to front, due to distorted reasoning.
> For nothing has been engendered in our body in order that
> we might be able to use it.
> It is the fact of its being engendered that creates its use.
> (5.832–35)

Lucretius certainly accepted end-directed thinking when it comes to human artifacts.

> Undoubtedly too the practice of resting the tired body
> is much more ancient than the spreading of soft beds;
> and the quenching of thirst came into being before cups.

Hence that these were devised for the sake of their use
is credible, because they were invented as a result of life's
experiences. (5.848–52)

It is just that he didn't want this analogy carried over to the living world. No values out there.

Quite different from these are all the things which were first
actually engendered, and gave rise to the preconception of
their usefulness later.
Primary in this class are, we can see, the senses and the
limbs.
Hence, I repeat, there is no way you can believe
that they were created for their function of utility.
(5.853–57)

Since he feels the need to warn us against it, Lucretius obviously recognizes that people think of organisms (or their parts) as having purposes. He is not prepared to deny that the world, the organic world in particular, shows design-like features. As an aside, therefore, treating the atomists in their own right and not just as a foil for Plato and Aristotle, perhaps rather than saying that atomists like Lucretius gave a heuristic understanding to purpose—something positive in the sense that it leads to insights, and that we shall see in later thinkers—it is more accurate to say that (outside human artifacts), they didn't really think it existed at all in reality (in the sense of having actual design or purpose) and only comes in as a sign of weakness in thinking. Either way, it is this approach that Plato (and almost certainly Socrates) thought improbable to the point of impossibility. No matter how infinite time and space may be, it isn't going to happen. To use a modern analogy, no number of monkeys randomly striking the keys of no number of typewriters is ever going to turn out the *Collected Works of William Shakespeare*, or to use a more contemporary example of the Roman orator and philosopher Cicero (106–43 BC), no number of letters of the alphabet shaken up in a bag are ever going to produce the *Annals* of Ennius, an epic poem about Roman history.

The World Soul

The context here is with living beings. As the discussion goes, Plato makes it clear that he is happy to extend forward-looking thinking to inanimate objects also; they too can be considered teleologically in terms of the designing intelligence deciding what is best for them. We can ask about purposes, as long as we can see value. Apparently, it would be perfectly proper to say that the earth is round rather than flat because it is in the middle of the universe, and that this is the best possible place for it to be. In other words, the earth is round in order that it might be in the middle of the universe. Unfortunately, in Plato's opinion, Anaxagoras, who has been noted as a forerunner in thinking about these sorts of things, gave up on the job and didn't really try to carry things through thoroughly. Having introduced the notion of end causes, he rather ignored them. In another dialogue, the *Timaeus*—very influential for this, or rather the first part, was virtually the only actual dialogue known to later generations until well into the Middle Ages—Plato himself took up the job and showed how it is that Mind orders everything for the best. Well known is the central claim of the *Timaeus* that the world—meaning the universe—is or was essentially disordered and then a designing mind, or Mind, imposed functioning order upon it. There is discussion about whether this Mind—what Plato called the Demiurge—was in fact a being who acted temporally, imposing its will upon an existing universe, one that had no beginning and will have no end. Or is it more a principle of ordering that always had its will impressed upon physical reality? Most of a philosophical vein have gone for the second interpretation, but there have been those (including Aristotle) opting for reading Plato as positing an actual creation. The Stoics (of whom more shortly) liked this idea for it tied in with their belief of eternal recurrence—worlds have beginnings and ends, and then start all over again.[11] No real matter to us. Either way, the Demiurge is a designer from the already-existing rather than a Creator from nothing, as is the Christian God.

The Demiurge is external to the world, but it goes one step further than perhaps necessary, for it imbues the world with a soul of its own. By "world," as is made clear by a later work, is meant the universe—it is quite false "that all the bodies that move across the heavens were mere collections of stone and earth."[12] By "soul" here (and elsewhere) Plato is not so much thinking of the Christian sort of soul, something purely mental and conscious—although that is certainly involved, especially intelligence—but also (as comes across clearly in *The Republic*, where Plato distinguishes but recognizes the appetitive part of the soul from the parts producing thinking and courage) something of the general life force that drives organisms forward. So, in other words, the universe is a living entity and ipso facto teleological thinking about its parts, both what we would normally judge the living and the nonliving, is not just appropriate but demanded: "the world is an intelligent being with its own soul, an arrangement ensuring that it is intelligently governed all the way down."[13] And value is right at the center of this. In *Candide*, Voltaire, through the mouthpiece of the philosopher Dr. Pangloss—who manages to get the clap and consequent rotting of virtually all of his bodily parts—pokes fun at the claim of Gottfried Leibniz (1646–1716) that this is the best of all possible worlds. The great German came by his thinking honestly, because this is at the heart of Plato's value-impregnated vision of the world. "Well, if this world of ours is beautiful and its craftsman good, then clearly he looked at the eternal model."[14] The reference here was to the Platonic Theory of Forms, most clearly discussed in *The Republic*. Deeply influenced by the School of Pythagoras, which combined a perhaps expected veneration of mathematics with a perhaps unexpected worship of the Sun, Plato argued that just as in this world we have physical objects (including organisms) that in some real sense owe their very being to the Sun, connecting all together through being the source of energy and also the power through which we can see the objects, so there is a world of rationality where we find ideal archetypes (Forms or Ideas) with the ultimate, the Form of the Good, analogously to the Sun connecting all together through giving the Forms their very being and enabling us through the intellect to know them.

It is this Form of the Good that was the Demiurge's guide. "Now surely it's clear to all that it was the eternal model he looked at, for, of all the things that have come to be, our universe is the most beautiful, and of causes the craftsman is the most excellent. This, then, is how it has come to be: it is a work of craft, modeled after that which is changeless and is grasped by a rational account, that is, by wisdom."[15] Notice the emergence of the major Greek theme, of the importance of wisdom, of rational thought. This is the sort of thing that made Tertullian very tense and is a constant worry in Christian (especially Protestant) thought, where the stress is on faith, something equally open to the untutored. The end for Plato, the ultimate value or values that make sense of our world of experience, could never be simply pig pleasures—food and drink and a nice wallow in the mud, literal or metaphorical. There is a hierarchy of values—of purposes—and wisdom, intelligence, and rationality are at the top. The Demiurge "put intelligence in soul, and soul in body, and so he constructed the universe."[16]

Aristotle

Aristotle, known as the "Stagirite" because of his origins, was, in the tradition of the best students, both follower and critic of his teacher Plato. He was a follower because with Plato he saw things in an organic mode, more so in fact because he was, for some period of his life, a hands-on biologist. Hence, like Plato, he was convinced absolutely and utterly that teleological thinking is not merely permissible but obligatory. "Nature never makes anything without purpose."[17] He was a critic because he denied the outside Mind or Designer or Demiurge and wanted nothing of world souls—especially not the latter, for reasons that we shall see are rather odd, taken on their own, but that make perfect sense within Aristotle's system. In other words, because he did not think teleology merely heuristic, although he certainly did not deny that side to teleology,[18] he was rather pushed toward our second option, that teleology is in some sense a function of, necessitated by, a kind of underlying end-directed force or forces—although we shall have to unpack the meaning of that. The main point is that when we talk of purpose, we are in some sense talking of something objective rather than just subjective, although, again, what that means precisely needs to be analyzed.[19]

Start with Aristotle's famous fourfold categorization of causes.[20] Take the making of a statue, let us say, of an unknown British soldier in the Great War. First there is (using this term in a slightly tighter sense than before) the "efficient cause"—the sculptor or modeler who made the statue. Then there is the "material cause"—the material out of which the statue was made, marble or bronze perhaps. Third, there is the "formal cause"—the actual shape of the statue. It would not be the soldier if you gave it four legs. You are almost certainly going to put it in uniform, and you are not going to put a German helmet on its head. And of great interest to us, the "final cause"—the reason you made or commissioned the statue, the purpose. Perhaps it was to mark the hundredth anniversary of the Battle of the Somme, which began on July 1, 1916, the dreadful day that saw 60,000 British casualties, including 20,000 dead. The battle that led to years of ill health and the eventual premature death of my paternal grandfather.

Great thinker that he was, Aristotle did not feel himself constrained by these categories. Formal causes, for instance, are much bound up with final causes. You are hardly going to mark the

Battle of the Somme if the form of your statue looks remarkably like Sophia Loren. In his extended discussion of animals, which is clearly at the heart of the very notion of teleological explanation, and where he shows his great concern with what I have termed the first part of the design argument—establishing the essentially purpose-like nature of organisms—Aristotle collapses the causes down to two: efficient and final. He speaks of things coming about by necessity, meaning that they are produced by efficient causes. He speaks also about things occurring for the sake of ends. Any adequate analysis must deal with both of these types of cause. "In dealing with respiration we must show that it takes place for such or such a final object; and we must also show that this and that part of the process is necessitated by this and that other stage of it."[21] Notice that, as always, value is the underlying theme. The very essence of showing the final object of respiration is that of demonstrating that the process is of value to the organism that is respiring.

Plato is a dualist with respect to the body-mind problem. Somewhat analogous to the later (seventeenth-century) philosopher René Descartes—of whom more later—Plato thought the body and the mind or soul are two separate entities or substances but interconnected, for the soul is not just the mind but the vital principle animating the body, and as such is in some sense located in the body. This, we shall see, was not Descartes's position, but there is overlap in the fact that mind can or should exist independently of the body—the main thrust of the *Phaedo* is to show how the soul can survive death. Aristotle's position is more subtle and complex.[22] It is his notion of form that does the heavy lifting here. Or to use a term that we shall see has had a long shelf life, "organization": "The body so described is a body which is organized."[23] Just as the shape of the soldier—its form—gives meaning to the statue, so the form of the organism is that which gives it life and meaning. It is that which gives the organism its soul—with the proviso that there are vegetative souls for plants, animated souls for animals, and intelligent souls for humans. But even at the vegetative level, organisms are not just bumps on a log. In his major discussion of the topic of soul, Aristotle calmly runs together the formal cause with the final cause: "It is manifest that soul is also the final cause. For nature, like thought, always does whatever it does for the sake of something, which something is its end."[24]

In the case of humans, we can see readily how we as organisms strive to our final causes. We have a goal and thanks to intelligence we can aim for it, even if we do not always achieve our goal. Our parts serve the ends of the whole, and the whole has clearly animal functions, like survival and reproduction, but also higher aims. In the cases of lower organisms, the parts likewise serve the ends of the whole and the whole likewise that of survival and reproduction. Is there any further aim, specifically, ends with respects to humans? Do cows exist for the purposes of humans? Does the moon? There is divided opinion on this, but in one sense at least, Aristotle is unambiguous about our importance. Apparently, we may infer "that, after the birth of animals, plants exist for their sake, and that the other animals exist for the sake of man . . . Now if nature makes nothing incomplete, and nothing in vain, the inference must be that she has made all animals for the sake of man."[25] As we shall see, this is not the end of the story.

What is clear is that, in the case of nonhumans and in the case of human parts, making allowance for human guidance through agriculture, the ends are achieved without thought and intelligence. Unlike Plato, there is no overall guiding, conscious design. That said, Aristotle does seem to appeal to some kind of vital force, an end-directed motivator, which raises the question: Why is it of general value for plants and animals to reproduce? Even if ultimately it is for our benefit, one looks for an immediate cause, something that gives them the drive, as it were, that is then going to benefit us. One suggestion Aristotle makes, which we shall see fits in with his overall world picture, is that in reproduction, although organisms do not become eternal, they get as close to the eternal as possible, and that in itself is a good. "The acts in which [the soul] manifests itself are reproduction and the use of food, because for any living thing that has reached its normal development . . . the most natural act is the production of another like itself, an animal producing an animal, a plant a plant, in order that, as far as nature allows, it may partake in the eternal and divine. That is the goal to which all things strive, that for the sake of which they do whatsoever their nature renders possible."[26]

It is important to get a handle on what is meant by "force" in such a context as this. There is a bit of a tendency to think of it rather like the background setting that makes the *Hound of the Baskervilles* such a terrifying story—a low-lying fog over the mire, a sort of semi-ethereal but physical thing enveloping everything. And it is true that this kind of idea can be found in some Aristotle-influenced systems. In chapter 8, we shall look at the French philosopher Henri Bergson (1859–1941), who spoke of some kind of life force that he called the "élan vital."[27] Suffice it to say here that this really does seem to have physical existence in some sense. It certainly has such an existence in the thinking of the novelist D. H. Lawrence, whose masterworks, the sequential novels *The Rainbow* (1915) and *Women in Love* (1921), are thoroughly infused with Bergsonian philosophy. Lawrence metaphorically translates the élan vital into blood, and again and again it functions as a kind of life force.

> Her warm breath playing, flying rhythmically over his ear, seemed to relax the tension. She could feel his body gradually relaxing a little, losing its terrifying, unnatural rigidity. Her hands clutched his limbs, his muscles, going over him spasmodically.
>
> The hot blood began to flow again through his veins, his limbs relaxed.
>
> "Turn round to me," she whispered, forlorn with insistence and triumph.
>
> So at last he was given again, warm and flexible. He turned and gathered her in his arms.[28]

To be repeated, many, many times!

This is not what is being supposed by Aristotle. For him, it is more a principle of ordering, not so much a thing but a relationship. Think of a right-angled triangle. For a Platonist, what is going on here is readily understandable. There is a Form, or something akin to a Form, that is a right-angled triangle, and triangles of this world conform to it—they "participate" in the Form. Likewise with purpose. For Plato, there is something out there, the Demiurge, which is itself bound up with the Form of the Good, something that is imposing order on the world. The Aristotelian is not a realist like this, nor is he or she a nominalist in thinking that it is all words or thought (by us) that confers its nature on the triangle or likewise the purposes of the world. The Aristotelian is not an eliminator. It is more a matter of a principle—that is why the link between the soul and the formal cause is so crucial. It is the form, in the Aristotelian sense, that gives the structure—and with it the meaning. The statue of the British "Tommy," a term used of the private soldier in the war, is not in itself the final cause, but by virtue of what it is, it points that way. Memory and commemoration. It is a candidate for our heartbroken respect. And note as always, there is value. We prize the statue because it points to an end that is good—memory of sacrifice and suffering, and determination to prevent its reoccurrence.

The World Picture

The living world is at the heart of Aristotelian end-directed thinking.[29] This comes through clearly in the *Physics*, shortly after he has introduced the fourfold causal division. "The necessity in nature, then, is plainly what we call by the name of matter, and the changes in it. Those causes must be stated by the student of nature, but especially the end; for that is the cause of the matter, not *vice versa*."[30] Can we be sure that there is this kind of necessity? Aristotle is eager to show that what is happening is not something that demands or involves conscious thought. "This is most obvious in the animals other than man: they make things neither by art nor after inquiry or deliberation . . . If then it is both by nature and for an end that the swallow makes its nest and the spider its web, and plants grow leaves for the sake of the fruit and send their roots down (not up) for the sake of nourishment, it is plain that this kind of cause is operative in things which come to be and are by nature."[31] So how do we analyze things? "And since nature is twofold, the matter and the form, of which the latter is the end, and since all the rest is for the sake of the end, the form must be the cause in the sense of that for the sake of which."[32] There is, incidentally, an almost proto-Aristotelian passage in Plato's *Philebus* that would fit in here:

> I hold that all ingredients, as well as all tools, and quite generally all materials, are always provided for the sake of some process of generation. I further hold that every process of generation in turn always takes place for the sake of some particular being, and that all generation taken together takes place for the sake of being as a whole.[33]

What of the rest of existence? Lower down, as it were, the material world does not have soul or souls. It couldn't have really. Aristotle subscribed (as did Plato) to the idea of the four elements: earth, water, air, and fire. (There is also the ether up above.) These elements just don't have the kind of integration leading to persistence and reproduction, the very things that we associate with organisms. This is far from saying that final-cause talk is inappropriate, apart from the form that might be imposed from without like the form of the statue. Aristotle (no less than Plato) is firmly committed to "global" or "cosmic" teleology. Right in the middle of his discussion of animal purpose, he stops to affirm that cosmological purpose is, if anything, more basic.

> Moreover, nothing abstract can be an object studied by physics, because nature does everything for a purpose. For just as

> in artifacts art is present, in things themselves there appears another such principle and cause, which, like the hot and the cold, we have from the universe. Hence it is more reasonable for the heaven to have come to be by the agency of such a cause, if it has come to be and to be because of such a cause, than for mortal animals.[34]

He explains that animals live and die, and change through time, whereas the heavens stay constant, and this reaffirms the centrality of the physical over the organic when it comes to final-cause thinking.

Aristotle saw a natural ordering of the elements, with earth at the center and then the others respectively moving outward. He also saw them as moving in some sense to reach their appointed places—this is where they should be. That is why heavy objects fall to the ground and smoke rises, why the oceans are above the seabed, and the winds above the water. They cannot stay at rest because of the actions of organisms and the disruptive effects of the heavens above, but they are directed to their proper places: "upward locomotion belongs naturally to fire and downward to earth, and the locomotions of the two are certainly contrary to each other."[35] This is why final-cause talk is appropriate. Although Aristotle is not naive. He is fully aware that it is at times proper to speak of things as being accidental or contingent. He doesn't think that an eclipse of the moon is necessarily for any great purpose. Is this just an exception to final-cause thinking? Not really. The eclipse as eclipse is not a substance. Heavenly beings move in circles because that is the perfect figure and so that is part of their nature. But the effects are not substances and so not necessarily explicable in terms of final cause. "Nor does matter belong to those things which exist by nature but are not substances; their substratum is the *substance*. E.g. what is the cause of eclipse? What is its matter? There is none; the *moon* is that which suffers eclipse. What is the moving cause which extinguished the light? The earth. The final cause perhaps does not exist."[36] Finally, the heavens themselves. Aristotle thought the stars were living beings, perpetually moving in circles, held in place by concentric, transparent, celestial spheres (the ether). "There is one heaven" only, that is ungenerated and eternal, and "its movement is regular."[37] This is the perfect motion and that in itself is reason enough for their existence. And so to the ultimate cause of it all, the Unmoved or Prime Mover. Beyond the sphere of the stars—not in the physical sense but in the metaphysical sense of having no place—exists the ultimate Being, that which is cause of itself and infinitely good. "The first mover, then, of necessity exists; and in so far as it is necessary, it is good, and in this sense a first principle."[38] It is that which motivates everything. "There is, then, something which is always moved with an unceasing motion, which is motion in a circle; and this is plain not in theory only, but in fact. Therefore, the first heavens must be eternal. There is therefore also something which moves them. And since that which is moved and moves is intermediate, there is a mover which moves without being moved, being eternal, substance, and actuality."[39] This explains the reproduction of plants and animals. They are becoming as close to the eternal as possible. So in an important way, Bergson and Lawrence are not completely out of line. They both sensed that what is at work here is something dynamic. It is not something static—2 + 2 = 4 was true, is true now, and will always be true, unchanging—but a real force, something organic, changing and striving toward an end. That is the real import of the blood metaphor. It is not something just existing, a substance, but something driving people forward, pushing them toward their goals. Although to be candid, I am not sure that what Lawrence had in mind was quite the state of intellectual ecstasy prized by Aristotle.

At times, Aristotle suggests that there are as many Unmoved Movers as there are celestial spheres, fifty-five in total number, but his general philosophy is that there is and can be only one primary mover. This Unmoved Mover is not a Creator God, as for the Jew or the Christian. Nor is it a designer God, the Demiurge of Plato. It is cause in the sense of final cause, that which is the end or goal for the cosmos as a whole: "it is vital to realize that the [Prime Mover] is a cause only in so far as it is the object of desire. It does not directly impart motion to the spheres; rather it excites in them the desire to emulate, in so far as they are capable of doing so, its state of pure intellectual activity."[40] In Aristotle's words: "And the object of desire and the object of thought move in this way; they move without being moved. The primary objects of desire and of thought are the same. For the apparent good is the object of appetite, and the real good is the primary object of wish."[41] How can this be, for almost paradoxically the Unmoved Mover is probably unaware of our existence? It is doing the only thing a perfect thing can do, namely, contemplate perfection, which means that by its very nature it is contemplating itself! "Are there not some things about which it is incredible that it should think? Evidently, then, it thinks that which is most divine and precious, and it does not change, for change would be change for the worse, and this would already be a movement."[42] Continuing: "Therefore it must be itself that thought thinks (since it is the most excellent of things), and its thinking is a thinking on thinking."[43]

On the surface, this is decidedly odd. I suspect that we have all known people who are pretty good at contemplating themselves and only themselves—I have one or two young relatives who have made an art form of this—and while we might regard them with amusement or irritation, we hardly regard them with respect. It follows naturally within the Aristotelian system and gives us good indication of how highly Aristotle, and the Greeks generally, regarded the life of pure thought, of ultimate rationality. It explains why there could be no world soul for Aristotle. He is prepared to think of the world organically, as an army, for instance. Asking about how the universe is good, he invokes the military. "For the good is found both in the order and in the leader; and more in the latter; for he does not depend on the order but it depends on him."[44] The household is another metaphor used.[45] Not a world soul, however, for this would mean that the Unmoved Mover was concerning itself with the well-being of physical things. Notice how, introducing the military metaphor, Aristotle stresses how order depends on the leader rather than the leader caring about the troops. The nature of the Unmoved Mover explains also why—perhaps almost callously to us who have been soused in other philosophies—Aristotle regards the highest aim of humankind to be that of rational reflection. Not something one would normally expect to occur (as it does) in the middle of a work on ethics. It is in this activity that we most directly aim to match the activity of the perfect being, the Unmoved Mover.

> If happiness is activity in accordance with virtue, it is reasonable that it should be in accordance with the highest virtue; and this will be that of the best thing in us. Whether it be reason or something else that is this element which is thought to be our natural ruler and guide and to take thought of things noble and divine, whether it be itself also divine or only the most divine element in us, the activity of this in accordance with its proper virtue will be perfect happiness. That this activity is contemplative we have already said.[46]

Of course, this doesn't mean that Aristotle is indifferent to our duties, our usual moral obligations, but that—as is also the case of Plato's philosopher kings—the most desirable activity is rational contemplation of the highest order of things. The philosopher kings know about the Forms and hence have the ability and obligation to run society, but power is never an end in itself. One relinquishes it as soon as one can. University administrators take note! Again, as with Plato, while the Greeks were not prudes or ascetics for the sake of prudishness or asceticism, for Aristotle, satisfying basic needs could never be the ultimate or primary aim of the good life. "Any chance person—even a slave—can enjoy the bodily pleasures no less than the best man; but no one assigns a slave a share in happiness—unless he assigns to him also a share in human life."[47]

The Stoics

The legacy is twofold. On the one hand, we have philosophical understanding of the nature of explanation based on the way we understand the world. Thanks to the Greeks, most people saw that forward-looking explanation, in terms of purpose or of final cause, seems not just appropriate but necessary. On the other hand, we have theological inference from this understanding to the existence of a deity or deities. Not all accepted this inference, but some did, and obviously, it is going to be of great importance as we move forward into the Christian era. But before this new religion established itself on the scene, the inference was embellished and promoted. Above all, it was the centerpiece of the thinking of the Stoics, a school started by Zeno of Citium (third century BC) and much favored by Roman thinkers, notably, the emperor Marcus Aurelius (AD 121–180). Cicero's *On the Nature of the Gods*, written in 45 BC, gives a good account of the Stoic position. There is a clear affirmation of the Designer at work, a wonderful anticipation of what we shall see in English thinking nearly two millennia later.

> When we see something moved by machinery, like an orrery [mechanical model of the solar system] or clock or many other such things, we do not doubt that these contrivances are the work of reason; when therefore we behold the whole compass of the heaven moving with revolutions of marvelous velocity and executing with perfect regularity the annual change of the seasons with absolute safety and security for all things, how can we doubt that all this is effected not merely by reason, but by a reason that is transcendent and divine?[48]

It is made clear that this applies not just to the world at large but particularly to the living world. "Why should I speak of the amount of rational design displayed in animals to secure the perpetual preservation of their kind? To begin with some are male and some female, a device of nature to perpetuate the species. Then parts of their bodies are most skillfully contrived to serve the purposes of procreation and of conception, and both male and female possess marvelous desires for copulation."[49] What is interesting is the explicit emphasis on ourselves. Of course, Plato and Aristotle were more interested in us humans than in other organisms, but the universe does not exist for us. We are part of the universe. In the *Laws*, Plato chides: "you perverse fellow . . . you forget that creation is not for your sake; rather you exist for the sake of the universe."[50] It is true that, within the world, Aristotle seems to privilege us, but overall we are out of luck: "if the argument be that man is the best of the animals, this makes no difference; for there are other things much more divine in their nature even than man, e.g., most conspicuously the bodies of which the heavens are framed."[51] The Unmoved Mover doesn't even know about us. For the Stoics, we are not perfect, but we are a very important part of the picture and there are implications that the designer had us humans in mind, in an exalted role. This is Aristotle pumped right up: "the corn and fruits produced by the earth were created for the sake of animals, and animals for the sake of man: for example the horse for riding, the ox for ploughing, the dog for hunting and keeping guard; man himself however came into existence for the purpose of contemplating and imitating the world; he is by no means perfect, but he is 'a small fragment of that which is perfect.' "[52]

Everything has purpose. Whether in Aristotle, certainly in the Stoics, we are special. And that, for good or ill, is surely an appropriate point from which to move on.

CHAPTER TWO

Jerusalem

WHEN JESUS DIED on the cross, when he rose from the dead, when he ascended into heaven, there was no Christian religion. There is good reason indeed to think that, until the last minute, Jesus did not know that he was going to die, and that the moment of realization and acceptance—"Father, into thy hands I commend my spirit" (Luke 23:46)—is the key to the whole drama. The work of making a religion from the life and teaching of Jesus fell to his followers, initially, Peter and Paul, and then over the next few centuries to the so-called church fathers, culminating with the greatest of them all, Saint Augustine of Hippo.

Confessions of a Neoplatonist

Augustine (354–430), whose influence on Western Christianity (Catholic and Protestant) cannot be overemphasized, was born in a Roman province in North Africa of a Christian mother (Monica) and a pagan father. Raised a Christian, he dropped out, acquired a mistress with whom he lived for thirteen years and by whom he had a son who died in adolescence, went to Italy as a professor of rhetoric, fell among the Manicheans (who believed in two gods, one good and one evil), sloughed off his first mistress and had another for two years, and then, finally back in

Africa, particularly at the urging of his very persistent mother, became again a Christian and was baptized by Saint Ambrose (ca. 340–397), bishop of Milan, in 386. In succession, Augustine became a (celibate) priest and then bishop of Hippo (in 395), and spent the rest of his life writing frenetically against a variety of Christian heretics. His Greek was bad, so he read the Bible and philosophy only in Latin, original or in translation. He knew of the works of the great philosophers Plato and Aristotle, but probably only read parts of the *Timaeus* and *Meno*, and got his knowledge of the rest from commentaries. The greatest influence, for all that he criticizes them, were the Neoplatonists, particularly Plotinus (204–270). Let us start there, with the Greek input to Augustine's thinking.

About ten years after returning to Christianity, Augustine wrote his autobiography, the *Confessions*, perhaps the greatest spiritual story of personal growth of Western culture. His God is emphatically the God of Plato, the God of *The Republic*, where the Form of the Good is a necessarily existing eternal force or entity, outside time and space, truly good and beautiful, the font of all other beings, from which everything stems and to which everything relates as the cause of existence. Unlike the Good that as Demiurge simply designs, the Christian God is the Creator from nothing. In all other respects, Augustine's deity is taken from the heart of the Platonic philosophy and then Christianized. Necessary: "For God's will is not a creature but is prior to the created order, since nothing would be created unless the Creator's will preceded it. Therefore God's will belongs to his very substance."[1] Outside space: "no physical entity existed before heaven and earth."[2] Outside time: "Your 'years' neither come nor go. Our years come and go so that all may come in succession. All your 'years' exist in simultaneity, because they do not change; those going away are not thrust out by those coming in . . . Your Today is eternity."[3]

One particularly innovative doctrine in Christianity is the Trinity. This is the teaching "that the Father, and the Son, and the Holy Spirit [Ghost] intimate a divine unity of one and the same substance in an indivisible equality; and therefore that they are not three Gods, but one God."[4] Plato in the *Timaeus* was ahead of things here, with the Demiurge doing the designing, the Forms being that which guided the Demiurge, and the world soul being that which gave life and meaning to the physical reality. Before Augustine, the fourth-century Christian, Calcidius, had already helpfully made the connections. The Demiurge, the Creator, is God the Father. The Good that gives rise to the Forms from which the Creator works and models the world is God the Son. And, neatly, the world spirit that pervades all physical reality is God the Holy Ghost. Plotinus (who never mentioned Christianity) was likewise helpful in explicating and unpacking the Form of the Good. He spoke of three aspects—"hypostases"—of this Form: the One, the Intellect, and the Soul. The One is "a nobler principle than anything we know as Being; fuller and greater; above reason, mind and feeling; conferring these powers, not to be confounded with them."[5] Absolutely crucial is the way that it is a unity, something entire and simple in itself, and giving integration to everything else. From it, all other things "emanate" (in the sense of being existence-dependent). Next comes the Intellect. This is bound up with the realm of the other Platonic Forms. As we know, they all are given their being by the Good, and it is these that the rational part of the soul can apprehend. As with Plato in *The Republic*, we get the analogy of the sun and its warmth and rays. "The Intellectual-Principle stands as the image of The One, firstly because there is a certain necessity that the first should have its offspring, carrying onward much of its quality, in other words that there be something in its likeness as the sun's rays tell of the sun."[6] And so third we come to the Soul. Basically, this is the world soul of the *Timaeus*. We have the Intellectual Principle setting the norms through the Forms. We have matter. And we have the Soul in some sense giving shape and meaning and motion to matter, in the light of the Forms.

In a Christianized form, these pagan ideas resonate through Augustine's writings. Not just the One, the Good, the eternal being from which all else emanates but also in the metaphors that are used to describe the other parts of the threefold nature of ultimate being. In his greatest work, *The City of God*, Augustine explicitly links Plato, via Plotinus, to the Trinity. Jesus is the Intellect, with the analogy of the sun and light. "This is in harmony with the Gospel, where we read: 'There was a man sent from God whose name was John; the same came for a witness to bear witness of that Light, that through Him all might believe. He was not the Light, but that He might bear witness of the Light. That was the Light which lighteth every man that cometh into the world.'"[7] And then there is the Soul. "Expounding Plato, Plotinus asserts, often and strongly, that not even the soul which the Platonists believe to be the soul of the world derives its blessedness from any other source than does our own soul: that is, from the light which is different from it, which created it, and by whose intelligible illumination the soul is intelligibly enlightened."[8]

What of purpose? What of value? What of the argument from design? Augustine's Christianity would not have been overly helpful here. There was King David's contribution, the opening of Psalm 19: "The heavens declare the glory of God; and the firmament sheweth his handiwork." Saint Paul also rushed briefly over the idea: "For the invisible things of him from the creation of the world are clearly seen, being understood by the things that are made, even his eternal power and Godhead; so that they are without excuse" (Rom. 1:20). Generally, however, in both Old and New Testaments, God's existence is a given, not something needing proof. The very idea that God might not exist is an unspoken-of nonstarter. Even the fool who "hath said in his heart there is no God" almost certainly was not declaring for atheism but for gods other than the God of the Jews, Yahweh.[9]

Augustine would have found more support in his Roman heritage. Noted already is the enthusiasm of the Stoics. Thanks to Cicero, we have seen that they gave a kind of protoversion of the argument as made famous at the beginning of the nineteenth century by Archdeacon William Paley, where the world of organisms is analogically linked to a functioning watch. But above all, it was the Platonic influence that was all-important here, as Augustine plunged headfirst into the teleological argument. The world shows signs of purpose at work. The only way we can explain this is through a good, creative, designing god or God. In the *Confessions*: "I was wholly certain that your invisible nature 'since the foundation of the world is understood from the things which are made, that is your eternal power and divinity' (Rom. 1:20)."[10] And then repeated in the *City of God*: "Even leaving aside the voices of the prophets, the world itself, by the perfect order of its changes and motions, by the great beauty of all things visible, claims by a kind of silent testimony of its own both that it has been created, and also that it could not have been made other than by a God ineffable and invisible in greatness, and ineffable and invisible in beauty."[11] We have a world of great value, created on purpose by a loving God.

History

From early in the *Confessions*: "Hear me God. (Ps. 54: 2). Alas for the sins of humanity! (Isa. 1:4) Man it is who says this, and you have pity on him, because you made him, and did not make sin in him. Who reminds me of the sin of my infancy? for 'none is pure from sin before you, not even an infant of one day upon the earth' (Job 14: 4–5)" (8–9). An infant of one day a sinner? "I was on the way to the underworld, bearing all the evils I had committed against you, against myself, and against others—sins both numerous and serious, in addition to the chain of original sin, by which 'in Adam we die' (1 Cor. 15:22)" (82). "The chain of original sin"? What is this? Why am I condemned? Why is the infant condemned? "You had not yet forgiven me in Christ for any of them, nor had he by his cross delivered me from the hostile disposition towards you which I had contracted by my sins" (82). Whatever else we might say, this is not the world of Plato and Aristotle.

Augustine drew heavily on Greek philosophy. But this comes in an altogether new framework. With Christianity, we have entered a world alien to and unknown by the Greeks. It is true that Augustine, thanks to his philosophy, strove for continuity, even as he changed Christianity to a nigh unrecognizable extent. Perhaps "created" is a better word than "changed." It is undeniable that we have another world that is going to give us a whole new take on purpose and on ends, a whole new take on values. This is a world that is in some very deep sense *historical* and that posed new thinking and challenges, indeed, to such a degree that, as Augustine changed what had been before, it in turn gave rise to a world picture that is foreign to anything remotely imagined by those early church fathers, including Augustine himself.

Start with history. When you first encounter Aristotle's philosophical writings, probably the thing that strikes you most forcibly is the staggering breadth of interest that he shows. He seems to write on just about everything—logic, mechanics, cosmology, optics, meteorology, biology, psychology, metaphysics, ethics, aesthetics, politics, rhetoric—you name it. Except history. He is no proto-Hegelian, giving us a story of world spirits through the ages and stuff like that. And it is easy to see why neither he nor Plato could really be philosophers of history. Their philosophical systems did not admit of such systems. They both saw the world as eternal. While Plato had the idea of the Demiurge, it may have been a principle of ordering, and even if it was not, it was working on stuff that had always been and always would be. The same for Aristotle. There were, of course, Greek histories, not to mention epic poems, but the background context is that such change is within limits and there are things before and things after. At most you get cycles. Plato's *Republic* spends the first half of the book building up the society, but at the end he talks of it decaying and collapsing—presumably to start all over again. Aristotle's

Unmoved Mover is eternal, indifferent to human beings, who generation after generation spend their lives trying to emulate it. With ends as far as individual humans are concerned, but not with ends over history. You are never going to become an Unmoved Mover.

The atomists are not into history in any meaningful sense, that is, history showing a pattern and perhaps a sense of purpose. Lucretius has a long discussion about the beginnings of civilization, but it all seems to be a matter of chance, not always for the good. Take religion. It stems from false impressions and misreadings, for instance, that there must be gods that drive the heavenly bodies. And what a disaster that all was. The New Atheists could not say it better.

> O humankind unhappy!—when it ascribed
> Unto divinities such awesome deeds,
> And coupled thereto rigours of fierce wrath!
> What groans did men on that sad day beget
> Even for themselves, and O what wounds for us,
> What tears for our children's children! Nor, O man,
> Is thy true piety in this: with head
> Under the veil, still to be seen to turn
> Fronting a stone, and ever to approach
> Unto all altars; nor so prone on earth
> Forward to fall, to spread upturned palms
> Before the shrines of gods, nor yet to dew
> Altars with profuse blood of four-foot beasts.[12]

Things happen, sometimes good things, like the coming and use of fire or the smelting of metals, and sometimes bad things, like religion. But it is all a matter of blind law.

Christianity is completely different. Thanks to the Jews, we have a clear start to things. "In the beginning God created the heaven and the earth." Then after putting the universe in order and filling it up, God brought the story to a climax by creating *Homo sapiens*. "God created man in his own image, in the image of God created he him; male and female created he them" (Gen. 1:27). As is well known, things rather went downhill from here. Placed in paradise, Adam and Eve were given but one prohibition: namely, not to eat fruit from the tree of good and evil. This, thanks to the seducing wiles of the serpent, they promptly did, and, equally, God promptly kicked them out of paradise, made them work for a living, and condemned them to eventual physical death. However, God took pity on humankind and so sent Jesus, his son—an essence (hypostasis) of himself—to be our savior. This was effected by death on the cross, which made possible our eternal salvation, escape from the life of sin, and reunion with God. Augustine is quite detailed on this, thinking it a kind of perpetual peace with the Almighty. He is also quite detailed on the fate of those who are not saved. There are going to be the ongoing torments of hellfire.

The disobedience of Adam and Eve—Augustine sees Adam as the real culprit—brings sin into the world, and this is something somehow passed down through the generations. It seems that the transmission of this dark aspect to our nature—"original sin"—is somehow bound up with the act of sexual intercourse. No doubt a reflection of his guilt over his misspent youth, Augustine sees even sex within marriage as in some sense a cause for shame. We have to do it: "Nevertheless, when that act is actually being performed, not even the children who have already been born from it are permitted to witness it. This right action desires recognition by the light of the mind, but it nonetheless shuns the light of the eye. Why is this, if not, because something which is by nature decent is performed in such a way as to be accompanied by shame, by way of punishment?"[13] Note that Christ escapes original sin because he was virgin born. For the rest of us, even newborn infants, we are tainted and bound to fall into moral error. The sacrifice of Jesus on the cross leads to the extension of mercy to some of us sinners. Let us give thanks for that, even though, unfortunately, it cannot be extended to the majority: "if all were to be brought across from darkness into light, the truth of retribution would have appeared in no one. But many more are left under the punishment than are redeemed from it, so that what was due to all may in this way be shown."[14]

Christian Purpose

Augustine is firmly committed to the traditional restricted number of years for the time span between the creation of the world and the coming of Jesus. He is significantly more hesitant about the time span from the death of Jesus to the coming Day of Judgment, but the overall impression is that it is going to be in the order of a thousand years. The important point is that with Christianity, unlike the Greek philosophies, we have a definite, temporally limited story, one with a beginning, a middle, and a predicted end. One that makes humankind the central players. One where the Creator God plays a definite and ongoing role. Which means that we have a whole new picture of direction, of ends, of purposes. We have a new picture of purpose or purposes for God, and we have a new picture of purpose or purposes for humans.

What, first, of God? What did he value? He did not have to create the world and its denizens up to and including humankind. Although, ultimately, Augustine would have insisted that we are not necessarily to know all of God's purposes, the main reasons are clear for all to see. God wanted creatures whom he could love—we are in an important sense God's children—and in turn, he wanted creatures who could love, worship, and obey him, who would give thanks for the gift of life and the opportunity to live it properly. As a Platonist, Augustine always insisted that evil is not a positive thing in its own right. The Forms come from the Good, so there could never be a Form of Evil. Plato even denied you could have Forms of hair and mud and dirt.[15] Evil must be a deficit, a lack of goodness. Hence, that we did not obey God is not God's fault. "Alas for the sins of humanity (Isa. 1:4). Man it is who says this, and you have pity on him, because you made him and did not make sin in him."[16] It was God's goodness that, like a parent, he picked up after us when we had made a mess.

What, second, of our purpose? In a sense, it is pretty straightforward. Drawing a distinction between the City of God (the realm of obedience to and fellowship with God) and the City of Man (the everyday realm in which humans spend their lives), it is attaining the former that is our goal, the end purpose to our lives and actions. "The New Testament clearly reveals what is veiled in the Old; that the one true God is to be worshiped not for the sake of those earthly and temporal goods which divine providence grants to good and evil man alike; but for the sake of eternal life and everlasting rewards, and the fellowship of the supernal City itself."[17]

Augustine fills in the details, in ways that have proven incredibly influential in the history of Christianity. We are all sinners, and there is nothing we can do to merit salvation. As we have seen, God being a just God is perfectly within his rights in turning from us and condemning us to eternal hell flames. But in his mercy, God is going to redeem some of us. Although how much is up to God. "He simply does not bestow his justifying mercy on some sinners . . . He decides who are not to be offered mercy by a standard of equity which is most secret and far removed from human powers of understanding."[18] What are we supposed to do now in order to gain such redemption or at least to get into the game and have a chance? Protestants, notably Luther and Calvin, turned to Saint Paul for guidance. No doubt reflecting the fact that he had done absolutely nothing to merit his own salvation, Paul dismissed good works and put everything on belief, on commitment, on faith. "For all have sinned, and come short of the glory of God" (Rom. 3:23). None of us are up to the moral mark. "Where is boasting then? It is excluded. By what law? of works? Nay: but by the law of faith. Therefore we conclude that a man is justified by faith without the deeds of the law" (Rom. 3:27–28). Justification by faith alone. *Sola fide.*

Augustine knew full well that this could not be the whole story. He was keenly aware of the preaching of Jesus. The Day of Judgment is coming, when Jesus will separate the sheep on his right hand from the goats on his left. Did you feed the hungry? Did you give drink to the thirsty? Did you welcome strangers? Did you clothe the naked? Did you visit the sick? Did you give solace to those in prison? If you did this, caring about others, then heaven awaits. "Verily I say unto you, Inasmuch as ye have done it unto one of the least of these my brethren, ye have done it unto me" (Matt. 25:40). A sentiment backed by other biblical passages. "What doth it profit, my brethren, though a man say he hath faith, and have not works? can faith save him?" Continuing, "Even so faith, if it hath not works, is dead, being alone" (James 2:14–17). Faith is important, but it is not all-important. Good works count too.

> Now, if the wicked man were to be saved by fire on account of his faith only, and if this is the way the statement of the blessed Paul should be understood—"But he himself shall be saved, yet so as by fire"—then faith without works would be sufficient to salvation. But then what the apostle James said would be false. And also false would be another statement of the same Paul himself: "Do not err," he says; "neither fornicators, nor idolaters, nor adulterers, nor the unmanly, nor homosexuals, nor thieves, nor the covetous, nor drunkards, nor revilers, nor extortioners, shall inherit the Kingdom of God."[19]

This is the point. If good works do not follow faith, then faith is not enough and probably not genuine. "Paul and James do not contradict each other: good works follow justification." Note that this does not mean that good works are payment for entry into the City of God. Nor are they just a mark that you have genuine faith. In some sense, the faith confers merit on them and God will take note. Talking of Abraham's faith in God being so great that he was ready to sacrifice Isaac on God's command: "If

Abraham had done it without right faith it would have profited him nothing, however noble the work was. On the other hand, if Abraham had been so complacent in his faith that, on hearing God's command to offer his son as a sacrificial victim, he had said to himself, 'No, I won't. But I believe that God will set me free, even if I ignore his orders,' his faith would have been a dead faith because it did not issue in right action, and it would have remained a barren, dried-up root that never produced fruit."[20]

Free Will?

Yet, can we reconcile the acts of humans with the sovereignty of God? In order for your acts to have any merit, for them to have genuine purpose, you must in some sense do these of your own free will. However, God knows exactly what is going to happen in the future, whether you will do the acts or not. You are "predestined" to do what you do. Is this not incompatible with free will? Augustine (as did those who followed him, notably, Calvin) denies this vehemently. "You would not necessarily compel a man to sin by foreknowing his sin. Your foreknowledge would not be the cause of his sin, though undoubtedly he would sin; otherwise you would not foreknow that this would happen. Therefore these two are not contradictory, your foreknowledge and someone else's free act. So too God compels no one to sin, though He foresees those who will sin by their own will."[21]

Augustine agrees that if (say) the stars were controlling our fate, then we would have no free will, and hence there would be no reason for praise or blame, no cause for eternal salvation or eternal damnation. This is not the usual case. Take an analogy. Suppose you have two students, Mary and Norman, about to take a mathematics examination. The better will receive a scholarship, the worse nothing. In the first scenario, Mary is hypnotized the night before and fed the right answers. Nothing happens to Norm. Mary does much better on the exam, winning the scholarship. The teacher knew this would happen but, although Mary gets the award, she deserves no credit. Nor does Norm merit scorn. This is analogous to fate determining our actions. Now, in the second scenario, nothing happens to either of the examinees the night before. However, Mary is much brighter than Norm and does much better on the exam, winning the scholarship. Again, the teacher knew this would be the case, but his foreknowledge was in no sense biasing the outcome, and so Mary is deserving of reward in a way that Norm is not.

This is how Augustine regards free will and predestination. God knows what is going to happen, but what does happen comes from within the human and hence merits praise or scorn. Note that Norm is not fed bad information or given a defective intelligence. He gets nothing. Analogously, in the good works case, the sinner is not fed a bad will but rather is inadequate—he or she is not fed everything to make a successful will. Of course, you could say that God determines the moral equivalent of Mary's 140 IQ and Norm's 90 IQ. So, in the end, God is not entirely off the hook. This does seem to be Augustine's position, at least with respect to angels, and this presumably applies to humans also. On the one hand, "we must believe that the holy angels were never without a good will: that is, the love of God." On the other hand, "the angels who, though created good, have nonetheless become evil, became so by their own will." How come? "The fallen angels, therefore, either received less of the grace of the divine love than those who remained steadfast in the same love; or, if both good and bad angels were created equal, then, while the latter fell by their evil will, the former were more amply aided by God."[22]

In the end, it is all God's business and God's decision. " 'Naked came I out of my mother's womb, and naked shall I return thither: the Lord gave, and the Lord hath taken away; as it pleased the Lord, so has it come to pass: blessed be the name of the Lord.' As a good servant, Job held the will of his Lord to be a great treasure in itself, through attendance on which his spirit would grow rich."[23] This is what Christian commitment is all about. "For faith is only faith when it waits in hope for what is not yet seen in substance."[24]

Aristotle Again

Augustine's legacy has lasted down to the present and still has bite. In the last decade, at one of the leading liberal arts colleges in America—Calvin College in Michigan—a member of the Department of Religion lost his job for doubting the total truth of Augustine's account of original sin.[25] For us now, the all-important point is the extent to which Augustine's world picture was thoroughly end-directed and totally value-laden. God acted throughout with purpose, in the way he made the world and in his plans for the world and its inhabitants after he had finished this work, most particularly, in the way in which he was prepared to intervene when his children went astray. Jesus did not die on the cross for nothing. He died that we might get eternal salvation. His purpose was our well-being. Our actions, our purposes, must be framed in the light of that. Nothing else matters.

This is not to say that time stood still. Augustine's was the most important word. It was not the final word. With the birth and rise of Islam—the Koran (or Quran) was delivered to Muhammad in the first half of the seventh century (609–632)—the intellectual and spiritual heart of the Western world moved to the Middle East and then to the countries that increasingly came under the rule of the Muslims. Until AD 1000 or a little later, religion, science, and philosophy became the province of the Arab world, until slowly it started to seep back into Europe and the Christian lands. There is much discussion about the reasons for the decline in Europe. The popular eighteenth and nineteenth centuries' claim that it was all the fault of Christianity, with the emphasis on faith and not on reason and evidence, is too simplistic on its own but probably is not entirely false. However, the legacy of the Greeks, particularly of Aristotle, did start to make inroads, often at first in Latin from an Arab translation and only slowly from the original Greek, as long-neglected manuscripts were discovered in libraries and as language abilities started to gain momentum. Gerard of Cremona (1114–87) went to Toledo, learned Arabic, and set up a veritable factory of translation. Included were Aristotle's *Physics*, *On the Heavens* (*De Caelo*), *Meteorology*, and *On Generation and Corruption*. Other works followed soon thereafter.[26]

Things did not always go smoothly. There were theological worries about Aristotle. The Platonic world soul is ultimately something imposed from without and remaining ontologically separate. Aristotle's metaphysics rather implied a spirit actually in matter itself. This was felt to run uncomfortably close to pantheism. But, overall, the Aristotelian world system and metaphysical approach was too powerful to be resisted, and, indeed, there were good reasons why Christians might find it very congenial. Think of the cosmology with the earth at the center, in a state of constant turmoil, as opposed to the heavens perfect and forever cycling changelessly and perfectly. Wasn't this exactly what our religion taught us, with the home of the Christian drama at the center, in a state of constant turmoil, as opposed to the heavens perfect and forever cycling changelessly and perfectly? Moreover, this could be fitted seamlessly with Christian natural theology (meaning knowledge of God through reason and evidence), as was shown in great detail with much sophistication by the greatest Catholic thinker of them all, Saint Thomas Aquinas (1225–74). As a good Christian, he thought that faith—revealed theology, meaning knowledge of God through divine intuition or authority—must always reign supreme and, if need be, trump reason. But this could be turned to advantage. Aquinas admitted, for instance, that as an Aristotelian he had nothing philosophically against an eternal universe. It was just that as a Christian, revelation told him otherwise. There was a beginning, and God is Creator God. Likewise with the soul. As an Aristotelian, Aquinas was drawn to monism, thinking that there are not (as Plato and, following him, Augustine assumed) separate substances, body and mind, but rather one entity, the bodily person whose soul is its form. Unfortunately, to a Christian this seemed to imply that when the body dies so also does the soul. Not so, said Aquinas. In some sense, the thinking part of the soul survives—after all, the Unmoved Mover has no physical body. "Therefore, the intellectual principle, which we call the mind or the intellect, has an operation in which the body does not share. Now only that which subsists in itself can have an operation in itself. . . . We must conclude, therefore, that the human soul, which is called intellect or mind, is something incorporeal and subsistent."[27] In any case, the Christian belief in the resurrection of the body reunites matter and form, body and soul.

Aquinas's five proofs of the existence of God are well known. Four are variants on the causal or cosmological argument, a version of which can be found in Augustine. "Of all visible things, the world is the greatest; of all invisible things, the greatest is God. But we see that the world exists, whereas we believe that God exists."[28] Note that this argument, which may have come from Plotinus, does not necessarily prove God as Creator. It is more God as sustainer. The world is a contingent thing. Why, therefore, does it exist? Because a necessary being stands behind it, sustaining it. The relationship is less one of efficient cause and more one, perhaps, of formal cause. This is certainly the way that things come across in Aquinas, as can be seen from his first proof, which centers in on motion. "It is certain and in accord with experience, that things on earth undergo change. Now, everything that is moved is moved by something; nothing, indeed, is changed, except it is changed to something which it is in potentiality. Moreover, anything moves in accordance with something actually existing; change itself, is nothing else than to bring forth something from potentiality into actuality."[29] Aquinas now argues for an Unmoved Mover, but notice that this cannot be an (Aristotelian) efficient cause because this would trap you into an unacceptable infinite regress, something Aquinas thought a contradiction in terms. "But this process cannot go on to infinity because there would not be any first mover, nor, because of this fact, anything else in motion, as the succeeding things would not move except because of what is moved by the first mover, just as a stick is not moved except through what is moved from the hand. Therefore it is necessary to go back to some first mover, which is itself moved by nothing—and this all men know as God."[30] This is not identical to Aristotle's Unmoved Mover; it is Creator, and it is keenly aware of our existence. In the context of the proof, it is not far from Aristotle's notion. It is at least a formal cause, if not a full-fledged final cause.

One finds the same style of thinking in Aquinas's treatment of the argument from design. At one level, if you like, he is bound (in the terms of this book) to be a Platonist, because he sees God standing behind everything. But the foreground is entirely Aristotelian.

> The fifth way is taken from the governance of the world. We see that things which lack knowledge, such as natural bodies, act for an end, and this is evident from their acting always, or nearly always, in the same way, so as to obtain the best result. Hence it is plain that they achieve their end, not fortuitously, but designedly. Now whatever lacks knowledge cannot move knowledge and intelligence; as the arrow is directed by the archer. Therefore, some intelligent being exists by whom all natural things are directed to their end; and this being we call God.[31]

Note that Augustine (in the Platonic tradition) and Aquinas (in the Aristotelian tradition) both include the inanimate world as well as the animate world in their argumentation. They would have thought it odd to do otherwise, for their God was the Creator God of the whole universe—a universe of purpose.

An Aristotelian World Picture

Things did not stand still. Through what is known as the "High" Middle Ages (1200–1450), the Aristotelian world picture was accepted, refined, moved forward. Sometimes such care had to be taken to stay onside with Christianity that one suspects the underlying philosophy was what is known as "instrumentalism," where theories are just taken as computing devices to yield predictions, as opposed to "realism," where theories are taken to denote what is really true out there in the world. For instance, Aristotelian cosmology demands that the heavens rotate around the earth once a day—light, dark, light, dark. Nicolas Oresme (ca. 1320–82) pointed out it would make things a lot easier to assume that the earth rotates once a day and that accounts for that portion of the heavenly motions. He took care to show, anticipating an argument of which Galileo was to make much three centuries later, that this did not mean that bodies dropped from on high would land farther back because the earth had moved forward. As we see on board ships when objects are dropped, it is all a matter of relative motion. Nevertheless, Oresme—whether from conviction or tactically—stepped back sharply from a realist interpretation of his hypothesis, quoting Psalm 92: "For God hath established the world, which shall not be moved." And that, apparently, was the end of that argument.

In other cases, however, advances were made that seem not to have been theologically objectionable, that made good sense within the system, and persisted, perhaps, even weathering the Scientific Revolution, if emerging in transformed ways on the other side. The "impetus" theory of Jean Buridan (1295–1358) is a case in point. On the Aristotelian system, physical objects have their natural places—earth, water, air, fire. If you let go of a stone, it falls in order to reach its proper place. The smoke from a fire rises, for the same reasons. Why then, if you throw a javelin, does it not immediately fall to the ground? You have passed on something akin to what today we would call "momentum." "It is because of this impetus that a stone moves on after the thrower has ceased moving it. But because of the resistance of the air (and also because of the gravity of the stone) which strives to move it in the opposite direction to the motion caused by the impetus, the latter will weaken all the time. Therefore the motion of the stone will be gradually slower, and finally the impetus is so diminished or destroyed that the gravity of the stone prevails and moves the stone towards its natural place."[32] This isn't exactly the modern notion of momentum, but it is certainly a legitimate forerunner.

Hints of what is to come: discoveries and ideas that were to destroy not just Aristotelian physics but Aristotelian metaphysics also. We are in a world filled with purpose—everything, rocks, plants and animals, humans, the sun and the moon, the planets and the stars. Aristotelian final causes explain all. The Christian God stands behind the world, but it is the "Philosopher's" theory that explains it. Not for much longer.

CHAPTER THREE

Machines

No Christian could ultimately escape the implications of the fact that Aristotle's cosmos knew no Jehovah. Christianity taught him to see it as a divine artifact, rather than as a self-contained organism. The universe was subject to God's laws; its regularities and harmonies were divinely planned; its uniformity was a result of providential design. The ultimate mystery resided in God rather than in Nature, which could thus, by successive steps, be seen not as a self-sufficient Whole but as a divinely organized machine in which was transacted the unique drama of the Fall and Redemption. If an omnipresent God was all spirit, it was all the more easy to think of the physical universe as all matter; the intelligences, spirits, and Forms of Aristotle were first debased and then abandoned as unnecessary in a universe that contained nothing but God, human souls, and matter.

—A. R. HALL, *THE SCIENTIFIC REVOLUTION, 1500–1800*[1]

The Scientific Revolution, that stupendous change in worldview, is usually dated from the publication of Nicolaus Copernicus's *De revolutionibus orbium coelestium* (*On the Revolutions of the Heavenly Spheres*) in 1543, the work that put the sun rather than the earth at the center of the universe—the change from the geocentric to the heliocentric worldview—to Isaac Newton's *Philosophiæ Naturalis Principia Mathematica* (*Mathematical Principles of Natural Philosophy*) in 1687, the work that gave the causal underpinnings of the whole system as developed over the previous one hundred and fifty years. Historian Rupert Hall (quoted in note 1 above) put his finger precisely on the real change that occurred in the revolution. It was not so much the physical theories, although these were massive and important. It was rather a change of metaphors or models—from that of an organism to that of a machine.[2] By the sixteenth century, machines were becoming ever more common and ever more sophisticated. It was natural therefore for people to start thinking of the world—the universe—as a machine, especially since some of the most elaborate of the new machines were astronomical clocks that had the planets and the sun and moon moving through the heavens, not by human force but by predestined contraptions. In a word, by clockwork! Referring specifically to a device built in the late sixteenth century, Robert Boyle (1627–91) was explicit: the world is "like a rare clock, such as may be that at Strasbourg, where all things are so skillfully contrived that the engine being once set a-moving, all things proceed according to the artificer's first design, and the motions of the little statues that at such hours perform these or those motions do not require (like those of puppets) the peculiar interposing of the artificer or any intelligent agent employed by him, but perform their functions on particular occasions by virtue of the general and primitive contrivance of the whole engine."[3]

Final Cause?

The great French thinker René Descartes (1596–1650) argued that ontologically God created two basic substances—*res extensa* and *res cogitans*, things extended and things thinking. The mark of the material world is that it has spatial dimensions. It is completely inert, unthinking, basic. Prima facie, Descartes adopted a version of the pre-Socratic atomist thinking, where material substance comprises "corpuscles" moving blindly according to unbroken law. In spirit his thinking was very different—the atomists accepted the void while necessarily Descartes denied it (because it has spatial dimensions and hence is substance), and they thought the atoms could not be broken apart, whereas his spatial substance is infinitely divisible. The mark of the spiritual world is that it has thought. It conversely has no physical dimensions. Rocks and planets, seas and rivers are *res extensa*. So are plants and so, notoriously, are animals. Angels are pure *res cogitans*. Humans likewise are thinking substance. Picking up on a thought to be found in the *City of God*, Descartes made it central to his philosophy. *Cogito ergo sum*. I think, therefore I am. Humans, however, are unique in that, as well as thinking substance, they are also material substance, connected via the pineal gland. Hence, Descartes was a dualist, like Plato, but unlike Plato in that for the Greek philosopher, the mind was clearly located in the body, the very point of Descartes's system was that mind could be nowhere spatially.

In Descartes's ("Cartesian") system, influential in its own right and representative of general thinking by the mid-seventeenth century, there simply was no place for Aristotelian final causes. The idea that matter itself has a kind of motive force, directed toward ends and hence incorporating values, was a contradiction in terms. Ends and values are precisely the sorts of things that *res extensa* cannot have. In any case, Descartes noted (perhaps somewhat disingenuously), one could never be quite sure what end God intended: "there is an infinitude of matter in His power, the causes of which transcend my knowledge; and this reason suffices to convince me that the species of cause termed final, finds no useful employment in physical [or natural] things; for it does not appear to me that I can without temerity seek to investigate the [inscrutable] ends of God."[4]

God! Descartes may have kicked final causes out of his science, but God was as important to the Frenchman as he was to Saint Augustine and Saint Thomas. It was God who guaranteed what Descartes referred to as "clear and distinct" ideas, the very foundations of his system of knowledge. Without God, an all-deceiving Evil Demon (introduced in the *First Meditation*) could be wrecking everything. This meant that, after the Scientific Revolution, purpose and value were far from gone. And why should they be gone? Even if final cause was no longer that helpful within the system—what end does the moon serve as it moves through the heavens?—the system overall, as God's artifact, had to be considered teleologically. "And God saw every thing that he had made, and, behold, *it was* very good" (Gen. 1:31). No one denied this. The universe generally and Planet Earth specifically are the place created by God for his favorites, made in his image, *Homo sapiens*. "And we know that all things work together for good to them that love God, to them who are the called according to *his* purpose" (Rom. 8:28).

This rather suggests that although after the Scientific Revolution there was no place for an Aristotelian take on purposes and values, at a somewhat generic level there was still place for, and need of, a Platonic take. The God of the *Timaeus*, the Divine Artificer. As it happens, we can go beyond the generic and become very specific, because another way of regarding the Scientific Revolution is as the triumph of Platonism over Aristotelianism! Start with Copernicus. From the beginning, everyone saw that his move to heliocentrism (sun-centered universe) was not something dictated by the evidence. He wasn't much into that sort of thing at all. He could plausibly have been influenced by Aristarchus of Samos (ca. 310–230), the "Copernicus of Antiquity," who had proposed a heliocentric world system. But there are deeper, earlier causes—the Pythagoreans, who were virtually sun worshippers and who had the earth and the sun going around some unseen central fire, and, of course, their follower, Plato, who made the sun so great a factor in his philosophical system. As the Form of the Good in the rational world is the foundation and sustaining cause of the other Forms, so the sun in the physical world is the foundation and sustaining cause of the objects of this world. And as the Good lets us know the

Forms through the intellect, so the sun lets us know the world's objects through vision. Copernicus was a fanatic.

> The Sun sits enthroned in the midst of all. In this surpassingly lovely temple, could this luminary be placed in any position which would better illuminate all at once. He is justly called the Lamp, the Mind, the Ruler of the Universe. Hermes Trismegistus named him the Visible God; Sophocles' Electra called him the All-Seeing. So the Sun sits as upon a royal throne, ruling the planets, his children, who circle about him.[5]

Kepler thought much the same way, for all that it was he who dethroned the circle from its privileged status as the perfect geometrical form that the heavens must obey. Try as one might, "by the highest right we return to the sun, who alone appears, by virtue of his dignity and power, suited for this motive duty and worthy to become the home of God himself, not to say the first mover."[6] This Platonism—obsession with the place of the sun, esoteric mathematical knowledge, insistence that things are governed by perfect figures or forms—is right there at the heart of Kepler's most modern-sounding work. Famous is the way in which he spaced the planets out from the sun according to measurements yielded by the five perfect solids, something of which Plato made much in the *Timaeus*. Don't think it is just chance that there are six and only six planets (including the earth)! The Great Geometer in the Sky knew what he was about. Perhaps less famous but as committed is the way in which Kepler argued that there is a Platonic world soul governing physical reality. "The view that there is some soul of the whole universe, directing the motions of the stars, the generation of the elements, the conservation of living creatures and plants, and finally the mutual sympathy of things above and below, is defended from the Pythagorean beliefs by Timaeus of Locri in Plato."[7] Having given a Christian blessing to this kind of speculation, Kepler explored in some detail the analogies between the functioning of the earth's soul and more familiar bodily workings, arguing that "as the body displays tears, mucus, and earwax, and also in places lymph from pustules on the face, so the Earth displays amber and bitumen; as the bladder pours out urine, so the mountains pour out rivers; as the body produces excrement of sulphurous odor and farts which can even be set on fire, so the Earth produces sulphur, subterranean fires, thunder, and lightning; and as blood is generated in the veins of an animate being, and with it sweat, which is thrust outside the body, so in the veins of the Earth are generated metals and fossils, and rainy vapor."[8]

Galileo, if less ebullient, was as committed as any to the Platonic insistence on the significance of mathematics. "That the Pythagoreans held the science of number in high esteem, and that Plato himself admired the human understanding and believed it to partake of divinity simply because it understood the nature of numbers, I know very well; nor am I far from being of the same opinion."[9] We have a Creator God, a Divine Artificer, separate from his creation, but structuring it according to his purposes and imbuing it with his values.

The Anglican Compromise

One should add that for the English particularly—not the Scots—this was a particularly happy state of affairs.[10] Roiled by religious controversy for much of the sixteenth century, under Queen Elizabeth a "compromise" was achieved. Steering a middle way between the authority of the Church, which was central to Catholicism, and the authority of the Bible, which was central to Protestantism, especially the Calvinism of John Knox and his coreligionists to the north of the border, the Anglican Church made much of natural theology—God as revealed through reason and especially the senses. Christianity was put on a nice, comfortable, empirical basis.

> There is a book, who runs may read,
> which heavenly truth imparts,

and all the lore its scholars need,
pure eyes and Christian hearts.

The works of God above, below,
within us and around,
are pages in that book, to shew
how God himself is found.[11]

Many parsons, secure and well provided for by lifetime benefices, found time hung heavily on their hands, especially if they could afford curates to do much of the donkey work. Hence, seeking to avoid the temptations of the turf or the bottle—or worse—they turned, often in a very professional way, to the study of nature. An ongoing hobby of beekeeping or of orchid growing was not only a happy way to fill the hours but could be justified theologically as study of God's creation.

It is perhaps little surprise that, reflecting the Renaissance discovery of antiquity and the new emphasis on language skills, one of the most significant British philosophical movements of the seventeenth century was so-called Cambridge Platonism. Henry More, its most influential member, was sympathetic to much of Descartes's thinking. However, he broke with the Frenchman (and sided with Plato) in thinking that mind or spirit has dimensions. He thought it existed in space, just as does matter. For this reason, More had no trouble with a physical vacuum. Spirit exists even if matter does not. "A substance incorporeal, but without Sense and Animadversion, pervading the whole matter of the universe, and exercising a plastical power therein according to the sundry predispositions and occasions in the parts it works upon, raising such Phaenomena in the World, by directing the parts of the Matter and their Motions, as cannot be resolved into meer Mechanical powers."[12] It is not that this planet of ours is an organism as such, but that in a way the whole of the universe is infused with life. Not necessarily in a conscious way—note in the passage just quoted the life force seems more vegetative than animal—but in a way that animates and moves brute matter.

All very comforting and probably quite influential. The Cartesians always critiqued the Newton system on the grounds that gravitational attraction relies on the quite unacceptable (and Aristotelian-like) notion of "action at a distance." For Descartes and his followers, one thing can only move another thing if they are touching or end points of a chain of touching things. The Newtonians defended their system on instrumental grounds. Whatever the ontology, the predictive power of their system was definitive. But that there was something slightly occult about gravity was undeniable. It is likely that Newton was reflecting the influence of his good Cantabrigian friend More, and relying on a kind of world soul to keep things moving along.[13] But, of course, all of this came with a price or consequence. God is in the world but he is not part of the world. The physical world is a lifeless machine. Kepler, for all his Platonism—or perhaps because of his Platonism—knew the score. As for Robert Boyle, the clock metaphor rules triumphant: "It is my goal to show that the celestial machine is not some kind of divine being but rather like a clock."[14] And what this means then is that, although talk of purpose and value has its place in philosophy and theology, as far as science is concerned, increasingly it was seen to be superfluous. Simply not part of the discussion. Not useful and, if anything, liable to mislead. God, purpose, value—these are out of the discussion. In the words of one of the most eminent historians of the Scientific Revolution, God had become "a retired engineer."[15]

There was a rift in the lute. Or, perhaps more appropriately, a fly in the ointment. Robert Boyle, a leading mechanist, saw clearly that organisms did not fit this nice, tight picture. As he wrote in his "Disquisition about the Final Causes of Natural Things," happily taking the opportunity to make a philosophical point while putting the boot into the French: "For there are some things in nature so curiously contrived, and so exquisitely fitted for certain operations and uses, that it seems little less than blindness in him, that acknowledges, with the Cartesians, a most wise Author of things, not to conclude, that, though they may have been designed for other (and perhaps higher) uses, yet they were designed for this use."[16] Boyle continued that supposing that "a man's eyes were made by chance, argues, that they need have no relation to a designing agent; and the use, that a man makes of them, may be either casual too, or at least may be an effect of his knowledge, not of nature's." Apart from anything else, this takes us from the chance to do science—the urge to dissect and to understand how the eye "is as exquisitely fitted to be an organ of sight, as the best artificer in the world could have framed a little engine, purposely and mainly designed for the use of seeing"—but it takes us away from the designing intelligence behind it.[17]

Boyle was being forced into playing a double game here. His stance supposedly is not something threatening to the mechanical position. It complements it! How can this be so? Boyle is distinguishing between acknowledging the use of final causes qua science and the inference qua theology from final causes to a designing god. First: "In the bodies of animals it is oftentimes allowable for a naturalist, from the manifest and apposite uses of the parts, to collect some of the particular ends, to which nature destinated them. And in some cases we may, from the known natures, as well as from the structure, of the parts, ground probable conjectures (both affirmative and negative) about the particular offices of the parts."[18] Then, the science finished, one can switch to theology: "It is rational, from the manifest fitness of some things to cosmical or animal ends or uses, to infer, that they were framed or ordained in reference thereunto by an intelligent and designing agent."[19] From a study in the realm of science, of what Boyle would call "contrivance," in the realm of science, to an inference about design—or rather Design—in the realm of theology.

No one was really deceived, nor did they want to be. John Ray was the most eminent of a line of "parson-naturalists," stretching from the seventeenth century well into the nineteenth century, who did their (biological) science, happy in the knowledge that this testified to the existence of the Creator, and to his great and good designing powers. Ray's *The Wisdom of God Manifested in the Works of the Creation* (1691) was an exemplar, containing sophisticated discussions of taxonomy (classification) that anticipate the work of Carl Linnaeus in the next century. No one, certainly not Boyle or Ray, was challenging the machine metaphor. It was just that when it came to organisms, it was felt that something more was needed—and that something more was, in a very Platonic fashion, the guiding hand of the Great Anglican up above.

Cutting Both Ways

Of course, one way in which you might get Platonic purpose out of science—or, more particularly, from hugging and enveloping and (some might say) constricting or confining science—would be to get rid of God altogether. At the least then you would no longer feel compelled to look for purpose and value when faced with some disgusting animal like a leech or a dung beetle. But that proved more difficult than you might think. If anyone should have been able to do it, it would have been the great Scottish skeptic David Hume (1711–76). And this indeed he set about to do with some vigor in his *Dialogues Concerning Natural Religion,* started in the 1750s but eventually published (anonymously) in 1779, shortly after the philosopher's death. He showed that the traditional argument from design—the argument of Plato and Augustine and Aquinas—is riddled with problems. On the one hand, who is to say that there is only one designer, and who moreover is to say that this designer got things right straight off? Our experience of complex entities is that usually this is a group effort, drawing on the experience of many attempts—

sometimes failures, sometimes successes—in the past. "But were this world ever so perfect a production, it must still remain uncertain, whether all the excellences of the work can justly be ascribed to the workman. If we survey a ship, what an exalted idea must we form of the ingenuity of the carpenter who framed so complicated, useful, and beautiful a machine? And what surprise must we feel, when we find him a stupid mechanic, who imitated others, and copied an art, which, through a long succession of ages, after multiplied trials, mistakes, corrections, deliberations, and controversies, had been gradually improving?"[20] And was it just one workman? "And what shadow of an argument . . . can you produce, from your hypothesis, to prove the unity of the Deity? A great number of men join in building a house or ship, in rearing a city, in framing a commonwealth; why may not several deities combine in contriving and framing a world?" The trouble is, of course, that you are reading in your conclusion—a unique, all-powerful deity—right into your premises and then thinking that you have discovered or proved something.

Even worse when, on the other hand, you turn to the nature of this deity. Hume was not the first to bring up the problem of evil. It is there in the thinking of Epicurus a century after Plato. "Is God willing to prevent evil, but not able? Then he is not omnipotent. Is he able, but not willing? Then he is malevolent. Is he both able and willing? Then whence cometh evil? Is he neither able nor willing? Then why call him God?"[21] However, Hume (explicitly acknowledging Epicurus) put the case as forcefully as anyone had done before or after. What of "racking pains" brought on by "gouts, gravels, megrims, toothaches, rheumatisms, where the injury to the animal machinery is either small or incurable?" It is all very well to stress the good side to things; there is a bad side also. "Mirth, laughter, play, frolic, seem gratuitous satisfactions, which have no further tendency: spleen, melancholy, discontent, superstition, are pains of the same nature. How then does the Divine benevolence display itself, in the sense of you Anthropomorphites?"[22] Acknowledging the existence of both moral evil (the evil brought about by human actions) and natural evil (the evil brought about by natural processes), Hume argued that neither is compatible with an all-loving God who is in control of things. The argument from design simply doesn't do what it is intended to do.

Yes, but . . . Right at the end of the *Dialogues*, Hume (through the spokesman who seems most closely to resemble his position) does a virtual U-turn. Perhaps there is a god—even a God—after all.

> That the works of Nature bear a great analogy to the productions of art, is evident; and according to all the rules of good reasoning, we ought to infer, if we argue at all concerning them, that their causes have a proportional analogy. But as there are also considerable differences, we have reason to suppose a proportional difference in the causes; and in particular, ought to attribute a much higher degree of power and energy to the supreme cause, than any we have ever observed in mankind. Here then the existence of a DEITY is plainly ascertained by reason: and if we make it a question, whether, on account of these analogies, we can properly call him a mind or intelligence, notwithstanding the vast difference which may reasonably be supposed between him and human minds; what is this but a mere verbal controversy?[23]

How should we take this passage? Is Hume in the end really a theist, believing in a God much like the Christian God? Is Hume a deist, believing in a God who is an unmoved mover, perhaps setting all in motion at the beginning, but now sitting back and letting nature unfold on its own? Is Hume an agnostic or skeptic, thinking that we simply cannot say whether or not there is a deity and, if there is, of what nature? Is Hume an outright atheist, denying absolutely the existence of God or gods? We can sidestep this issue. The fact is that at the end of the *Dialogues*, Hume does qualify his arguments. One possible reason that strikes me as plausible is that although the argument from design is being presented as an argument from analogy—artifacts show the marks of design and do in fact have a designer; the world seems designed in the same way as artifacts; hence, by analogy, the world must have a designer or Designer—truly it is what Charles Sanders Peirce called an "abductive" argument and what today is often labeled "an argument or inference to the best explanation."[24] In *The Sign of the Four*, Sherlock Holmes nailed it in his explanation to Dr. Watson. "How often have I said to you that when you have eliminated the impossible, whatever remains, *however improbable*, must be the truth?" The point is that the organized complexity that we see in organisms particularly has to have some explanation and—pace the atomists—pure chance will not do the job. Hence, there must be a designer, and since we know that the designer was not human, there must be a God. Of course, this line of argument only works until a new and more successful challenger comes along; but until this happens, the conclusion reigns supreme.

It is quite possible that this is a major reason why, twenty-five years after Hume's work was published, the textbook writer Archdeacon William Paley was able to write and publish his hugely successful *Natural Theology*, a work that actually mentions the *Dialogues*, and yet with the most famous and, in respects, most influential positive exposition of the Design Argument.

> In crossing a heath, suppose I pitched my foot against a *stone*, and were asked how the stone came to be there; I might possibly answer, that, for any thing I knew to the contrary, it had lain there for ever: nor would it perhaps be very easy to show the absurdity of this answer. But suppose I had found a *watch* upon the ground, and it should be inquired how the watch happened to be in that place; I should hardly think of the answer which I had before given, that, for any thing I knew, the watch might have always been there.[25]

The watch shows organization, marks of design. The stone does not. Hence, there has to be a God. Shall we simply say that the watch just happened? "Or shall it, instead of this, all at once turn us round to an opposite conclusion, viz. that no art or skill whatever has been concerned in the business, although all other evidences of art and skill remain as they were, and this last and supreme piece of art be now added to the rest? Can this be maintained without absurdity? Yet this is atheism."[26]

One should add that there was undoubtedly a social element to all of this. Remember that natural theology for the Anglican Church represented the via media, the middle way between the extremes of Roman Catholicism and extreme Protestantism, notably Calvinism. The Church was —still is—part of the governing fabric of England, with leaders (bishops) members of the legislative body. The end of the eighteenth century was a tense time in Britain, with the awful example across the Channel of the French Revolution and then the rise of Napoleon and nigh twenty years of ongoing warfare. God—the warm, friendly God of the Church of England—was needed to maintain and justify stability. This held right through the middle of the nineteenth century—when the teaching of Paley at the older universities was at its peak.

> The rich man in his castle,
> The poor man at his gate,
> God made them high and lowly,
> And ordered their estate.[27]

Not just the nineteenth century. In the late 1940s, the infant lungs of Michael Ruse used to bellow out those words in assembly in (the state-run) Whitehall Primary School in Walsall, Staffordshire.

Immanuel Kant

Was there any way forward? Aristotelian final causes had been removed from the physical sciences. Purpose was gone. As Hume said, it was no longer acceptable to move from statements about matters of fact to statements about matters of value. Too often he found "that instead of the usual copulations of propositions, *is*, and *is not*, I meet with no proposition that is not connected with an *ought*, or an *ought not*." As he continued: "This change is imperceptible; but is however, of the last consequence. For as this *ought*, or *ought not*, expresses some new relation or affirmation, 'tis necessary that it should be observed and explained; and at the same time that a reason should be given, for what seems altogether inconceivable, how this new relation can be a deduction from others, which are entirely different from it."[28] Science is science and values are values, and yet that simply didn't seem to be true when we turn to organisms.

If anyone could extract us from this conundrum it would be the greatest philosopher of modern times, the late-eighteenth-century German Immanuel Kant (1724–1804). The key to his thinking lies, as it does for so many of us, in his childhood, and in particular in his being raised in a Lutheran Pietist family. Kant grew up in an atmosphere of intense Protestant spirituality, where the Bible was the guide and faith was the foundation. This in itself was going to make him wary of natural theology—the great Reformers Luther and Calvin were very suspicious of reason—and the arguments of people like Hume finished the job. One then sets this against the other great driving passion in Kant's intellectual life, his total confidence in Newtonian mechanics and the belief that we now can understand the nature of the physical world. Realizing that he had in some way to rise above the skeptical philosophical thinking of Hume, this led Kant to his critical philosophy, and in particular to his great leap forward—his philosophical Copernican Revolution—in seeing that our understanding of the world comes from within as well as without. Hence the synthetic a priori, necessary thinking for all rational beings, something Kant extended from science to morality also.

Hence also Kant's adamant insistence that knowledge, Newtonian physics, and proper moral thinking—guided by the Categorical Imperative—can only go so far. It cannot get us to God.

It might point that way, but genuine knowledge is impossible. The arguments of natural theology not only fail but were bound to fail. "I had to deny knowledge in order to make room for faith."[29] Note that for Kant this does not make God's existence any less secure or immediate or important. In fact, God is supremely important in understanding morality. For instance, the Categorical Imperative imposes an absolute ban on lie-telling. How can one justify this when some lies—what you might say to the Gestapo searching for rebels or a child dying of cancer—are surely morally demanded? Only by putting it within the context of God, who will make all right in the end. "If the strictest obedience to moral laws is to be considered the cause of the ushering in of the highest good (as upshot), then, since humans can't bring about happiness in the world proportionate to worthiness to be happy, an omnipotent moral being must be postulated as ruler of the world, under whose care this proportion is achieved. That is, morality leads inevitably to religion."[30]

A nice, neat solution until Kant turned to biology in the second half of the *Third Critique*, *The Critique of the Power of Judgment* (1790). Influenced by the biology of his day, and one very strongly suspects the philosophy of Aristotle, Kant came right up against the problem of purpose, of final cause. Kant is writing post–Scientific Revolution, so he wants nothing to do with final causes in physics nor does he (qua science) want anything to do with general ends. He is with the Dutch philosopher Baruch Spinoza (1632–77) on this: "There is no need to show at length, that nature has no particular goal in view, and that final causes are mere human figments."[31] God is a retired engineer. Kant did see (as did Aristotle) that not everything to do with organisms demands a teleological analysis. Grass grows but not in order to feed animals, although it is true that they take advantage of the grown grass. However, when it comes to organisms, Kant sees that they do seem to be organized, and that this organization leads to a kind of functioning—survival and reproduction. This means in some sense that the parts of organisms are both cause and effect, with the kind of forward-looking, value-impregnated dimension that one expects in a world of purpose. The eye, for instance, brings about survival and reproduction, which in turn brings about another eye. But this seems to take us beyond the machine metaphor. "In a watch one part is the instrument for the motion of another, but one wheel is not the efficient cause for the production of the other: one part is certainly present for the sake of the other but not because of it. Hence the producing cause of the watch and its form is not contained in the nature (of this matter), but outside of it, in a being that can act in accordance with an idea of a whole that is possible through its causality."[32] Kant goes on to say that it is a matter of organization or even self-organization. "This principle, or its definition, states: *An organized product of nature is that in which everything is an end and reciprocally a means as well.* Nothing in it is in vain, purposeless, or to be ascribed to a blind mechanism of nature."[33]

At one level, it seems that Kant is introducing an Aristotelian force of some kind. "An organized being is thus not a mere machine, for that has only a motive power, while the organized being possesses in itself a formative power, and indeed one that it communicates to the matter, which does not have it (it organizes the latter): thus it has a self-propagating formative power, which cannot be explained through the capacity for movement alone (that is, mechanism)."[34] We have seen that for Aristotle the formative power is not a thing like a mist above a swamp—it is more a principle of organization. Hegel thought the buck stopped here. Kant took over an Aristotelian position as against a Platonic position. "By means of the notion of Inner Design Kant has resuscitated the Idea in general and particularly the idea of life. Aristotle's definition of life virtually implies inner design, and is thus far in advance of the notion of design in modern Teleology, which had in view finite and outward design only."[35] Perhaps. Kant certainly agrees with Aristotle that teleology is uneliminable. You cannot go the way of the atomists. But Kant doesn't see that the Aristotelian ontological given is really allowable in the Newtonian world. Hence, barred as he is from the external teleology of Plato (faith cannot be knowledge), and barred as he is from the internal teleology of Aristotle (Newtonian physics does not allow principles making for final causes), Kant is driven to the third alternative: the teleology of biology is heuristic.

> The concept of a thing as in itself a natural end is therefore not a constitutive concept of the understanding or of reason, but it can still be a regulative concept for the reflecting power of judgment, for guiding research into objects of this kind and thinking over their highest ground in accordance with a remote analogy with our own causality in accordance with ends; not, of course, for the sake of knowledge of nature or of its original ground, but rather for the sake of the very same practical faculty of reason in us in analogy with which we consider the cause of that purposiveness.[36]

Kant is caught in a difficult bind. Of course, he thinks God is responsible for all of this. Given his underlying philosophy and theology, he cannot bring God into the scientific discussion. He cannot opt for an Aristotelian solution making final causes in some sense real. But he realizes that you cannot do biology without final-cause thinking. So the best he can do is say that teleology is a guide, a heuristic. He cannot say why, ultimately, we need it, but there we are. Although it doesn't stop Kant from being rather nasty about biology. You want to make the life sciences equal to the physical sciences? Fuhgeddaboudit! "[W]e can boldly say that it would be absurd for humans even to make such an attempt or to hope that there may yet arise a Newton who could make comprehensible even the generation of a blade of grass according to natural laws that no intention has ordered; rather, we must absolutely deny this insight to human beings."[37]

We enter the nineteenth century with a more fully articulated third option for purpose. Plato: God put purpose into the world—external teleology. Aristotle: purpose is part of the fabric of the world—internal teleology. Kant: purpose is heuristic, needed to do science but in itself of no ontological content—mind-given teleology. Obviously, in an important sense, this is in the tradition of the atomists. There is no teleology actually out there in the living world. Equally obvious, as just noted, it is not in the tradition of the atomists. Teleology is not just a product of sloppy or weak thinking. It is, in some sense, essential, and from this follows its vital heuristic nature. We cannot do without it,[38] which, in a way, flips us out of the atomist frying pan and into the Kantian fire. For the atomists there is nothing to explain because there is nothing. For Kant, there is a problem. We want to know why this appeal to ends is needed in biology and not physics. And that is about as good as we can get for the moment—which, as we shall see, turned out to be quite a good moment. Kant's insistence on the need of final-cause thinking in the biological domain bore heavy fruit. But we get ahead of ourselves.

CHAPTER FOUR

Evolution

THE GREEKS did not have a philosophy of history, something that turned them to thinking about purpose over long periods of time. That was Christianity's major contribution to the discussion. Augustine was as influential on the Protestants as he was on the Catholics—arguments could be made for saying that in many respects, Luther and Calvin were closer to him than was Aquinas. However, by the beginning of the eighteenth century, things were starting to come apart at the seams. Although both Catholics and Protestants were deeply committed to Christianity, the fact is that they did differ on important things, structurally (like the authority of the pope) and theologically (like the status of the Virgin Mary). With such fundamental differences, the way was open for those who wondered whether any of it was true. It did not help that the Augustinian view of God, akin to the Form of the Good, was in direct conflict to the biblical view of God endorsed by the Reformers. For them, God was a person—the stern judge of Genesis, the loving father of the parable of the prodigal son. This did not sound much like an ethereal being, outside time and space.

In parallel, voyages of discovery were going farther and farther afield, especially around Africa to the East. There Europeans found old civilizations, with their own religions, none of which had any role for Jesus of Nazareth. Who was right and who was wrong? The British rather compounded things for they hated and barred Christian missionaries, thinking (with good reason) that they only foment strife and are bad for trade. The trouble is that, if on commercial grounds you start making arguments about the integrity of Hinduism and Buddhism, before long you are liable to believe what you are saying. Science likewise has uncomfortable ways of pressuring Christianity. If we are now but one speck of dust in an infinite universe, no longer at the center, what price the special status of humans? Biblical time was also under pressure. With geological knowledge being increasingly important for mining and canal building and the like, the complex strata of the earth's crust hinted strongly at eons before the present, as did the fossils being unearthed. Then, too, the Christian religion itself was starting to crumble under new, sophisticated scrutiny. Spinoza was one of the first to start looking at the Bible as a humanly written book rather than given from on high by the Almighty. And truly, when you think about it, the legacy of Augustine is pretty dreadful. God is going to condemn most of his creatures to everlasting hellfire, and while he can justify this on the grounds that they themselves sinned and so are guilty, the reason for their troubles is that God didn't give them the wherewithal to stand firm against temptation.

Progress

This was not a reason to give up on history. But, as God was being expelled from the sciences, perhaps the time was now coming when God could be expelled from history also, or at least pushed to the sidelines. Could we move from a "Providential" view of history—where God controlled everything and nothing we humans did had any merit in its own right—to a more secular view of history where we humans, through our labors, could indeed have influence over our fate, over our future? Many in the eighteenth century, the Age of the Enlightenment, thought this indeed a possibility.[1] Thought and hope were actualized in the form of a formidable challenger: progress! No less end-directed, this was a philosophy of history that took the responsibility and control away from God and put it firmly in our hands. The values are our values. The purposes are ours and ours alone. A function both of the loosening grip of the God hypothesis and a growing sense that thanks to technology and political reform and more, it was now possible—and desirable—for us ourselves to make a better tomorrow. In the words of its most distinguished historian: "The idea of human Progress, then, is a theory which involves a synthesis of the past and a prophecy of the future. It is based on an interpretation of history, which regards men as slowly advancing—*pedetemtim progredietes*—in a definite and desirable direction, and infers that this progress will continue indefinitely."[2]

To be honest, the two philosophies (Providence and progress) were frequently not all that different, and at times it is difficult to distinguish their ends. If anyone can tell in the following passage where Kant stood, they are doing better than I: "Now if things in the world, as dependent beings as far as their existence is concerned, need a supreme cause acting in accordance with ends, then the human being is the final end of creation, for without him the chain of ends subordinated to one another would not be completely grounded; and only in the human being, although in him only as a subject of morality, is unconditional legislation with regard to ends to be found, which therefore makes him alone capable of being a final end, to which the whole of nature is teleologically subordinated."[3] Everyone was a millennialist, worrying about end times, and not too careful to distinguish spiritual heaven up above from secular heaven down here.[4] That said, there really was a change of attitude—for the progressionist, science, education, and political economy were crucially important in the New World scheme, and it was for us to use these and to push for better times.

Toward the end of the century, the essayist and novelist William Godwin (1756–1836)—husband of the feminist Mary Wollstonecraft and father of Mary Shelley, author of *Frankenstein*—was a great enthusiast. He had a nigh-fanatical belief in human perfectibility and consequent progressive improvement of society. He believed all of our weaknesses and moral failings could be overcome. This apparently was a function of our receptivity to truth. "Every truth that is capable of being communicated is capable of being brought home to the conviction of the mind. Every principle which can be brought home to the conviction of the mind will infallibly produce a correspondent effect upon the conduct."[5] And from the individual, because all humans are fundamentally the same, this will spin out to society. "We are partakers of a common nature, and the same causes that contribute to the benefit of one will contribute to the benefit of another."[6]

Expectedly, one sees variations across countries and across cultures. Great Britain was now a united country and getting well into the Industrial Revolution. Naturally, thoughts of progress reflected this. Adam Smith was important here, with his ideas of the importance of a division of labor and of the Invisible Hand making a virtue of individual selfishness. "It is not from the benevolence of the butcher, the brewer, or the baker that we expect our dinner, but from their regard to their own interest."[7] In France, until the revolution, intellectuals and other would-be reformers labored under the restraints of the Church and the monarchy, the ancien régime. One tended therefore to see more theoretical and idealistic arguments, as well as vitriolic attacks on the clergy and others in power. Hume may have hammered at the argument from design, but Voltaire was brutal about natural theology in his satirical *Candide* (1759), and in *The Nun* (1796), Denis Diderot was positively cruel about religion and its organizations.

Evolution

What one also sees is the linking of the cultural and the political with other areas of inquiry, for instance, about humankind. Naturally enough, thoughts of progress tended to get caught up with the comparative anthropology that was becoming increasingly detailed and comprehensive as Europeans extended their travels more and more broadly. It was not just a matter of improving our own society from where we are now but also of showing how far we—that is to say, we white people jammed in between the Atlantic and Asia—had already come. Diderot, a novelist and one of the founders of the *Encyclopédie* (a very Enlightenment attempt to register and catalog all knowledge), was forward but typical. "The Tahitian is at a primary stage in the development of the world, the European is at its old age. The interval separating us is greater than that between the new-born child and the decrepit old man."[8] Note the analogy with human individual growth. Naturally, this soon led to thoughts of what we would call organic "evolution." (Back then, "evolution" as a word tended to apply to individual growth.)

Evolution was not an idea entirely unknown. Empedocles, like the atomists, had protoevolutionary views—the kind of views, as we have seen, later endorsed by Lucretius. The elements come together randomly and sometimes form parts of animals and plants—a head here and a leg there—"Here sprang up many faces without necks, arms wandered without shoulders, unattached, and eyes strayed alone, in need of foreheads."[9] In turn these sometimes combine: "Many creatures were born with faces and breasts on both sides, man-faced ox-progeny, while others again sprang forth as ox-headed offspring of man, creatures compounded partly of male, partly of the nature of female, and fitted with shadowy parts." Every now and again, as Lucretius acknowledged, we get functioning organisms.[10] As we have seen, the philosophers—Aristotle picked out Empedocles—knew that this was silly talk. Randomness and chance do not make for functioning complexity. Evolution in any sense as we might understand it is simply physically impossible. This would apply to the physical world, but also very particularly to the world of animals and plants. Final causes demand something extra.

This kind of critical thinking persisted right down through our time period. In the *Third Critique*, Kant (somewhat tensely) argued that blind law doesn't lead to purpose, to phenomena demanding final-cause understanding. The "archaeologist of nature" can speculate all he likes, but ultimately he has to find in nature "an organization purposively aimed at all these creatures, for otherwise the possibility of the purposive form of the products of the animal and vegetable kingdoms cannot be conceived at all. In that case, however, he has merely put off the explanation, and cannot presume to have made the generation of those two kingdoms independent from the condition of final causes."[11] Kant's great follower, the French comparative anatomist Georges Cuvier, made much the same point. He stressed that in considering an organism, we have to look at how the various parts fit and work together. We have to dig into the organization of the organism and ask about purposes. Justifying this, as it were, was something Cuvier called the "conditions of existence." This demands that we look at the parts of organisms from a final-cause perspective.

> As nothing can exist without the re-union of those conditions which render its existence possible, the component parts of each being must be so arranged as to render possible the whole being, not only with regard to itself but to its surrounding relations. The analysis of these conditions frequently conducts us to general laws, as certain as those that are derived from calculation or experiment.[12]

We must keep value questions in front of us all the time. What is the purpose of a particular part? And from this it follows that any organism midway between two functioning organisms, which there would have to be if evolution be true, would be literally neither fish nor fowl and hence nonviable.

Yet increasingly, evolution was an idea whose time had come. As Kant admitted candidly, there were some very suggestive phenomena. Paradoxically, it was the philosophers who suggested that perhaps all organisms are related. Aristotle particularly took note of the isomorphisms—what we today call homologies—between members of different species. Could this mean something? In his rather convoluted way, Kant thought it might.

> The agreement of so many genera of animals in a certain common schema, which seems to lie at the basis not only of their skeletal structure but also of the arrangement of their other parts, and by which a remarkable simplicity of basic design has been able to produce such a great variety of species by the shortening of one part and the elongation of another, by the involution of this part and the evolution of another, allows the mind at least a weak ray of hope that something may be accomplished here with the principle of the mechanism of nature, without which there can be no natural science at all.[13]

Truly, however, it was progress that was the chief motive force and helped people to ride roughshod over problems with final cause. As with human societies, and likewise drawing analogy with the growth of the individual, Diderot wrote: "Just as in the animal and vegetable kingdoms, an individual begins, so to speak, grows, subsists, decays and passes away, could it not be the same with the whole species?" Going on to say that an organism might continue to exist "but in a form, and with faculties, quite different from those observed in it at this moment of time."[14] Cut from the same cloth, across the Channel, the Scottish-educated English physician and poet Erasmus Darwin (grandfather of Charles) held forth.

> Organic Life beneath the shoreless waves
> Was born and nurs'd in Ocean's pearly caves;
> First forms minute, unseen by spheric glass,

Move on the mud, or pierce the watery mass;
These, as successive generations bloom,
New powers acquire, and larger limbs assume;
Whence countless groups of vegetation spring,
And breathing realms of fin, and feet, and wing.

Thus the tall Oak, the giant of the wood,
Which bears Britannia's thunders on the flood;
The Whale, unmeasured monster of the main,
The lordly Lion, monarch of the plain,
The Eagle soaring in the realms of air,
Whose eye undazzled drinks the solar glare,
Imperious man, who rules the bestial crowd,
Of language, reason, and reflection proud,
With brow erect who scorns this earthy sod,
And styles himself the image of his God;
Arose from rudiments of form and sense,
An embryon point, or microscopic ens![15]

Biological progress, from the blob to the human, is a given, and Erasmus Darwin explicitly tied his biology into his philosophy. The idea of organic progressive evolution "is analogous to the improving excellence observable in every part of the creation; such as the progressive increase of the wisdom and happiness of its inhabitants."[16]

As the fortunes of cultural progress rose and fell, so rose and fell the fortunes of evolution. The French Revolution and the consequent Napoleonic wars made people very wary of happy stories about a better future. It is clear that for Cuvier, as for many English conservatives, his enthusiasm for final cause had an added political dimension. A consummate civil servant, he had lived through the horrors of the French Revolution, and he wanted nothing to do with dangerous ideologies, especially those stemming from science. Evolution was not just theoretically wrong—based on the identity between today's organisms and those mummified by the Egyptians, he would have added that it was empirically wrong too—it was politically dangerous as well. But as the nineteenth century calmed down and took speed, the Industrial Revolution caught fire again, especially with the building of the railways, and social progress came back into fashion. Evolution was not far behind. In his *Vestiges of the Natural History of Creation* (1844), the successful Scottish publisher of magazines for the general public, Robert Chambers, showed the way, arguing that nature reaches ever higher, and that it is progress that provides the needed proof, if need there be.

> A progression resembling development may be traced in human nature, both in the individual and in large groups of men. . . . Now all of this is in conformity with what we have seen of the progress of organic creation. It seems but the minute hand of a watch, of which the hour hand is the transition from species to species. Knowing what we do of that latter transition, the possibility of a decided and general retrogression of the highest species towards a meaner type is scarce admissible, but a forward movement seems anything but unlikely.[17]

What of times to come? "Is our race but the initial of the grand crowning type? Are there yet to be species superior to us in organization, purer in feeling, more powerful in device and act, and who shall take a rule over us!"[18] Chambers was confident: "There may then be occasion for a nobler type of humanity, which shall complete the zoological circle on this planet, and realize some of the dreams of the purest spirits of the present race."[19]

At midcentury all of this was picked up and made a smashing success by the most famous and defining poet of the Victorian era, Alfred Tennyson, in his lament for a long-dead friend. Perhaps, suggested Tennyson inventively in *In Memoriam* (1850), that friend was too advanced to live. Perhaps this planet was not yet ready for "a nobler type of humanity."

A soul shall strike from out the vast
    And strike his being into bounds,

And moved thro' life of lower phase,
    Result in man, be born and think,
    And act and love, a closer link
Betwixt us and the crowning race . . .

Whereof the man, that with me trod
    This planet, was a noble type
    Appearing ere the times were ripe,
That friend of mine who lives in God.

Purpose and Evolution

What price purpose in all of this? Overall, in these early years of its existence, evolutionary theorizing didn't really rise above the status of a pseudoscience.[20] People could see only too clearly that it existed on the back of what many (with reason) considered the very iffy ideology of cultural progress. One mark was the way in which nonprofessionals like Robert Chambers felt free to plunge right in with their ideas, as though they had spent their lives working in the laboratory or out in the field. And this really showed when it came to purpose. The leading professional biologist to get tangled up with ideas of evolution was the French naturalist Jean Baptiste de Lamarck, who published his speculations in his *Philosophie Zoologique* in 1809. That he was an enthusiast for cultural progress is shown if only by the fact that, although a minor aristocrat, it was during the revolution that his career really took off. He became a world-leading invertebrate taxonomist, a scientist of deserved respect, and as such was brought right up against the issue of the end-directed nature of the features of organisms. Famously, he spoke to this issue through the mechanism that now bears his name, the inheritance of acquired characteristics. Why does the giraffe have a long neck? Because its ancestors stretched up to eat the leaves from tall trees, and over the generations the necks became ever longer through this strenuous use. Conversely, cave dwellers are often blind simply because they never use their organs of sight.

As it happens, Lamarck was not the first to use this mechanism, and it was never his chief mechanism of evolutionary change (of which more later), but it does show that he was sensitive to issues of purpose at the individual level and spoke to them,[21] unlike others. To be fair, Erasmus Darwin did offer a bit of a hotchpotch of suggestions, included in which was what came to be known as "Lamarckism." This is more than one can really say for Chambers, for whom purpose at the individual level was never really a meaningful issue. That was not where his mind was at. Not that this meant he was indifferent to purpose. Always, historical purpose or end direction. Indeed, with his obsession about social or cultural progress, the very raison d'être of his thinking about organic evolution was progress from the primitive—in fact, Chambers was much into the spontaneous generation of life from nonlife—to the complex, or what was known back then as from the monad to the man. As was everyone else, from Diderot through Erasmus Darwin (whose poetry is explicitly on this topic) to Tennyson (whose poetry is no less explicitly on this topic). You want value? Evolution gave it to you.

What was the cause of historical direction? Like many intellectuals at the end of the eighteenth century—including many of the signers of the American Declaration of Independence—Erasmus Darwin was a deist, believing in God as Unmoved Mover. Unlike Aristotle's God, this was more an Unmoved Mover as efficient cause than as final cause, so if we were speaking generically, this would put him in the Plato camp. Probably the same can be said of Chambers, and certainly the same can be said of Tennyson, although he was quite explicitly an Anglican theist, so his chief influence was (and always was to be) Christianity. Interestingly, at a more specific level, as an educated Englishman, Tennyson knew his Plato, and in *In Memoriam* makes much use of the allegory of the cave—about seeing only indistinctly—taken from *The Republic*. Tennyson was comfortable philosophically as well as theologically with the idea of a Creator God who stands behind his world, and whose purpose was the nature and well-being of his favored creatures, humans.

The Perils of Purpose

From the perspective of our story, with the midcentury publication of *In Memoriam*, we have arrived at a fascinating point of tension. Purpose was no less important for the evolutionists than it was for their critics, and yet—thanks in no small part to the evolutionists, if only for the aspects of reality to which they were drawing attention—purpose was all over the place and, like Humpty Dumpty, it didn't seem that all of the king's men and all of the king's horses could put it together again. First, thoughtful people—especially those firmly within the scientific community, like Cuvier, or (initially) those commenting on the work of the scientific community, like Tennyson's teacher, the English historian and philosopher of science William Whewell (a scientist in his own right, especially with his work on the tides)—were adamant that the clue to understanding organisms was precisely that stressed by Kant.[22] Organisms had to be understood in terms of final cause. The parts of organisms did not exist in their own right. They existed in order to complete the whole, to serve the ends of the whole, which was survival and reproduction: "each part is conceived as if it exists only through all the others, thus as if existing for the sake of the others and on account of the whole, i.e., as an instrument (organ)."[23]

Second, there was recognition of the widespread existence of the already-mentioned isomorphisms, what by 1850 we can without anachronism call "homologies," between very different species of organism.[24] A much-favored example was the ordering of the bones in the forelimb of humans (used for grasping), of the wings of birds (for flying), of horses (for running), and of dolphins (used for swimming). This seems to serve no direct purpose. Does this mean that the world was not so very teleological after all? The trouble was that it was growing increasingly difficult to ignore homology. On the one hand, for all that Cuvier stressed the conditions of existence, it was homology that was the working tool of the comparative anatomist. No one thought that Cuvier could really in isolation deduce the nature of an organism from a bone. It was always done comparatively against other better-known organisms. Whether or not one was an evolutionist, one was interested in relationships even if only ideal. The English anatomist Richard Owen, drawing on Continental thinking, made much of the "vertebrate archetype," a kind of Platonic form—he identified it as such—that was a template for all vertebrates, including humans.[25]

On the other hand, there were theological virtues to recognizing homology. In the 1830s, thanks to a bequest by the Earl of Bridgewater, a series of works (eight in all) on natural theology were commissioned, written, and published.[26] Whewell drew the task of writing on cosmology, and this set him to thinking about purpose through the universe.[27] As a good Christian and Newtonian, the Aristotelian solution of heavenly bodies as living beings was obviously unacceptable, but what purpose did so great a creation actually serve? Could these heavenly bodies be themselves the homes of living beings? This hypothesis of a "plurality of worlds" had a venerable history of several centuries. But Whewell thought it raised immeasurable difficulties for the Christian. If the denizens of other worlds are not human or humanlike, then what is the point of them? If they are humanlike, does this mean that some at least are fallen and that the Savior has to go and care for them? Could Jesus be crucified over and over again throughout time and space?

How could Whewell wriggle out of this one? Two moves came at once to mind. In both cases: *Aegrescit medendo*. The cure is worse than the disease. First, one argues for what is sometimes known as the "argument (for God's existence) from law," that you really don't need useful ends—what are sometimes called "utilitarian" ends—to show design. Any kind of law-bound pattern will do. "[I]n the plan of creation, we have a profusion of examples, where similar visible structures do not answer a similar purpose; where, so far as we can see, the structure answers no purpose in many cases; but exists, as we may say, for the sake of similarity: the similarity being a general Law, the result, it would seem, of a creative energy, which is wider in its operation than the particular purpose."[28] Second, one might stress how very badly the world is designed anyway! So what if the planets and stars are useless? Much of our world is useless. Hundreds of organisms are born that wither and die without success. "Of the vegetable seeds which are produced, what an infinitely small proportion ever grow into plants! Of animal ova, how exceedingly few become animals, in proportion to those that do not; and that are wasted, if this be waste!"[29] Just as well, because there would be nowhere for them to live anyway. Huge areas of our planet are arid and dry and worthless. "Vast desert tracts exist in Africa and in Asia, where the barren land nourishes neither animal nor vegetable life."[30] Suddenly, all of those Kantian demands about purpose seem a lot less pressing and interesting. The ends are much broader and ecumenical than anyone might have thought, and in any case, lots of times there don't seem to be many ends in the first place.

These are the tangles of natural theology. On top of all of this there is the problem of purpose in what one might call a more positive sense. As Kant agreed, all of that talk about homology was very conducive to thinking favorably about evolution. Right at the end of his life, when he read (in translation) an evolutionary work (*Zoonomia*) by Erasmus Darwin, he might have become an evolutionist.[31] Even in the *Third Critique* he wrote of it as "a daring adventure of reason" and that "there may be few, even among the sharpest researchers into nature, who have not occasionally entertained it."[32] This said, though the evolutionary hypothesis is not "absurd," how do you speak to the purposeful nature of organisms? Blind law, let us agree, will not do the trick.

That just leads to mess and disorder. Unfortunately—or perhaps fortunately—even if we agree with Whewell that there is widespread mess and disorder (hardly something compatible with law-bound patterns), not everything is mess and disorder. The seeing eye is more than this. Or the thinking brain. The one attempt to speak to the effectiveness of organisms is the Lamarckian process of the inheritance of acquired characteristics. As Charles Darwin saw, this could hardly be the whole story. When he first became an evolutionist, he was more or less exclusively a Lamarckian, but he soon saw that it was not enough. "Wax of Ear, bitter perhaps to prevent insects lodging there, now these exquisite adaptations can hardly be accounted for by my method of breeding there must be some core[r]elation, but the whole mechanism is so beautiful."[33] You can ignore final cause. That was more or less the tactic taken by Chambers, that one suspects as much from ignorance as anything thought through carefully. But like cancer, it has a nasty way of catching up with you. And if you go the route of Tennyson and give the whole job over to God, you are really no longer doing science. Whewell following Kant was explicit about this. He felt that the problem of final cause meant that there could be no solution to the problem of organic origins. With mixed feelings—a plus for theology, a minus for science—he saw at once what this meant. "Science says nothing, but she points upwards."[34]

Teleology, final cause, purpose, seems to have collapsed in on itself. We need it but it is wreathed in paradox. Where to go next? Fortunately, the route was being opened by a newcomer to the field, the just-mentioned Charles Robert Darwin (1809–82). Let us turn to him and his work.

CHAPTER FIVE

Charles Darwin

CHARLES DARWIN, grandson of Erasmus Darwin, came from a rich, upper-middle-class family.[1] His other grandfather, also the grandfather of his wife Emma, was Josiah Wedgwood, who founded the pottery works that to this day bears his name. Already this gives us major clues into the nature and achievements of the younger Darwin.[2] On the one hand, he was not about to repudiate his heritage. Why would he? On the other hand, his heritage was precisely that of British industry and political economy and more. The other set of major clues into the nature and achievements of the younger Darwin come from his education and training. It is pretty clear that Darwin's father was an atheist and his mother's family, the Wedgwoods, were Unitarians, and hence, deists in major respects. However, Charles Darwin was given a conventional Anglican education, public school (meaning, really, private school), and then (after an unhappy interlude at Edinburgh, aiming to be a physician) Cambridge. No surprise here because such training was the sine qua non for entry into the level of society to which Darwin belonged. While an undergraduate, he had a full dose of Paley's natural theology, not to mention the influence of older members of the university, like Whewell, with whom he became intimate. Although he did not take a science degree—there were no such degrees—these older members were, to a person, ardent Newtonians, so informally, Darwin had good instruction in that direction. He was a lifelong amateur in the sense that the family money meant he never had to work for a living. He was a lifelong professional in the sense that he always knew the standards demanded of quality science.

Darwin spent five years (1831–36) as ship's naturalist to HMS *Beagle*, spending much time in South America and eventually going all the way around the globe. During this time Darwin's religious beliefs changed from fairly conservative Anglican to deist, a view he held for the next several decades, changing again at the end of his life to a form of agnosticism. Although by the nature of his work he had to spend much time thinking and writing about the science-religion relationship, he always claimed that by nature he was not a particularly religious man, and that is probably true. Darwin returned to England and in the next two years became first an evolutionist and then a Darwinian, meaning he discovered his mechanism of change, natural selection. What spurred the move to evolution was, above all, the distribution of the animals (birds, lizards, tortoises) on the Galapagos Archipelago, a group of islands in the Pacific that the *Beagle* visited in the final part of its journey. Why should they be similar but different, on islands even within sight of each other? Could it be that ancestors came, and as their offspring moved from island to island, they changed? What spurred the move to natural selection was the strongly felt need to be (the possibility of which Kant denied) the Newton of biology—to find a cause for the change.

For reasons that are still not entirely clear, for twenty years Darwin sat on his ideas until spurred into print by the arrival of an essay with much the same thinking, sent from the Far East by a young naturalist, Alfred Russel Wallace. And so finally *On the Origin of Species by Means of Natural Selection, or the Preservation of Favoured Races in the Struggle for Life* appeared late in 1859. The fat was in the fire, for by then Darwin was a well-known figure in Victorian culture.[3] His account of the *Beagle* voyage had found its way into many drawing rooms, and he was much respected both as an explorer and writer, but also because it was known that for many years he had been laboring on detailed investigations. Indeed, a long-term study of barnacles was the subject of (friendly) fun in a new novel by a best-selling author.

The Origin of Species

The *Origin* is a deceptively sophisticated piece of work.[4] The heart is the derivation of natural selection, called in later editions the "survival of the fittest." First Darwin argued to the "struggle for existence." This is a term introduced by the Anglican parson Thomas Robert Malthus at the end of the eighteenth century. Worried at what he thought were naive claims about social progress, Malthus argued that humans are always subject to the stern laws of population growth.[5] Numbers are always pressing to outgrow supplies of food and space. There will hence be a struggle for existence. Darwin generalized this to all of nature—in itself no big thing because, before focusing on humans, Malthus started with a general claim made by Benjamin Franklin.

> A struggle for existence inevitably follows from the high rate at which all organic beings tend to increase. Every being, which during its natural lifetime produces several eggs or seeds, must suffer destruction during some period of its life, and during some season or occasional year, otherwise, on the principle of geometrical increase, its numbers would quickly become so inordinately great that no country could support the product. Hence, as more individuals are produced than can possibly survive, there must in every case be a struggle for existence, either one individual with another of the same species, or with the individuals of distinct species, or with the physical conditions of life.[6]

Shortly we will get to issues of purpose and ends, but note how Darwin is seizing on the point that Whewell introduces and uses. Most organisms simply don't make it through. For Darwin this is going to be a central and, if the term is not inappropriate, positive thing. Not making it through is a crucial part of the picture. Whewell notes it and makes use of it, but as we have seen, in the process rather shooting himself in the foot. To save his revealed theology—our special relationship with Jesus Christ—he has to crimp his natural theology: many things don't have much purpose. We have a classic example here of what Thomas Kuhn argued was a crucial feature in scientific revolutions—it isn't just that the new theory (paradigm) has virtues, but that the old theory (paradigm) is collapsing in on itself.[7]

From the struggle, Darwin is now ready to move to selection. Constant, heritable variation seems always to be coming into populations; Darwin didn't have much clue about the reasons, but he was sure that it was there, and in the struggle, some variations are going to help their possessors to success. The variations do not appear to order, nor do they have direction—they are "random"—but they do make a difference. Hence, there will be what evolutionists today call a "differential reproduction."

> Let it be borne in mind how infinitely complex and close-fitting are the mutual relations of all organic beings to each other and to their physical conditions of life. Can it, then, be thought improbable, seeing that variations useful to man have undoubtedly occurred, that other variations useful in some way to each being in the great and complex battle of life, should sometimes occur in the course of thousands of generations? If such do occur, can we doubt (remembering that many more individuals are born than can possibly survive) that individuals having any advantage, however slight, over others, would have the best chance of surviving and of procreating their kind? On the other hand, we may feel sure that any variation in the least degree injurious would be rigidly

> destroyed. This preservation of favourable variations and the rejection of injurious variations, I call Natural Selection.[8]

Continuing to hold for now on discussion of purposes and ends, in the *Origin* Darwin embedded this mechanism in a larger picture or framework. As Newton used his force of gravitational attraction to explain a range of phenomena, from the motions of projectiles down here on Earth to the paths of the planets in the heavens, so Darwin used his force of natural selection to explain a range of phenomena. He ran the gamut, from social behavior, through paleontology and the fossil record, on to geographical distribution—pride of place was given to those inhabitants of the Galapagos. He moved on through classification and morphology—all of those homologies came out here—and so to development and why the embryos of organisms, very different as adults, nevertheless have embryos that are virtually indistinguishable. With vestigial and useless organs, like the appendix, Darwin was done. He had given a paradigm case of what Whewell called a "consilience of inductions."[9] Many areas of scientific inquiry were gathered together beneath one overall causal hypothesis. In turn, this overall causal hypothesis threw explanatory light on many areas of science.

The Descent of Man

In the *Origin*, Darwin deliberately stayed away from our species. He wanted to get his general theory on the table first, as it were. But he always did think his theory applied to our species—in his private notebooks in 1838 just after he had discovered natural selection, the first reference to the new mechanism is to us, and not just to us but to our brains and capacity for thinking. Hence, lest he be accused of cowardice, right at the end of the *Origin* he wrote of us. "Light will be thrown on the origin of man and his history." But no one was fooled, and almost immediately Darwin's work was known as the "monkey theory." I am not sure that he would ever have written a full-length work on humans had it not been for the apostasy of Wallace who, in the 1860s, became enamored of spiritualism and started arguing that human evolution demands divine intervention. Horrified, Darwin again put pen to paper and in 1871 published *The Descent of Man and of Selection with Relation to Sex.*

Much in the *Descent* is on familiar ground, applying natural selection to our physical features and showing the links with other animals, especially the apes. There are extended discussions of religion and morality, topics to which we shall return. The really innovative part of the book—so innovative that it causes imbalance, taking up well over half of the pages—is a long discussion of a secondary mechanism that Darwin introduced into the *Origin*, sexual selection. This occurs within a species and is for mates, coming in two forms: male combat, where males fight it out for access to the females, as when rutting stags clash antlers, and female choice, as when the peahens choose the peacock with the most flamboyant tail feathers. There was a simple reason for this new emphasis. Wallace argued that natural selection had to be inadequate because many human features, like our hairlessness, have no direct adaptive function. There had to be another cause, or as one might say, Cause. Darwin agreed with Wallace about the inadequacy of natural selection but argued that sexual selection could do the job.[10] Something like hairlessness was all a matter of taste. The less apelike you looked, the more the girls would fancy you.

Purpose

Now let's meet some of our promises. Darwin used to joke that he could have transcribed Paley's *Natural Theology* by heart.[11] Except he wasn't joking. From his earliest readings as a schoolboy, along with the overriding importance of the machine metaphor, he was drenched in the argument from design. From a chemistry textbook that he and his brother used for amateur experiments in the back garden: "The animal body may be regarded as a living machine, that obeys the same laws of motion as are daily exemplified in the production of human art."[12] At the same time, we learn that we have "a body of incontrovertible evidence of the wisdom and beneficence of the Deity."[13] At Cambridge, Darwin became an avid beetle collector and got more of the same. The standard work on British insects was the *Introduction to Entomology* by the Reverend William Kirby and William Spence.[14] Expectedly, one learns that looking at and collecting insects is no purely secular activity. For good Anglicans, it is akin to being in church: "no study affords a fairer opportunity of leading the young mind by a natural and pleasing path to the great truths of Religion, and of impressing it with the liveliest ideas of the power, wisdom, and goodness of the Creator."[15]

And so it goes. Darwin knew all about homology. During his years in Edinburgh, he worked with the anatomist Robert Grant, an evolutionist who would have filled his head with such themes. Again, after the *Beagle* voyage, just when he was at his creative peak, he became friends with the young Richard Owen, who was thinking through his ideas about archetypes. Notwithstanding, it was the design-like nature of organic ends that grabbed his imagination. Why wouldn't it? Cuvier's *Le règne animal distribué d'après son organisation pour servir de base a l'histoire naturelle des animaux* was in the *Beagle* library, and even Darwin, with his Englishman's blockage about the languages of others, could make out: "L'histoire naturelle a cependant aussi un principe rationel qui lui est particulier, et qu'elle emploie avec avantage en beaucoup d'occasions; c'est celui *des conditions d'existence*, vulgairement nommé *des causes finales*."[16] Then at Cambridge, Whewell and the other scientists, like John Henslow the botanist and Adam Sedgwick the geologist, hammered in the theme at every opportunity. They had to, because they themselves were under threat from more conservative thinkers who were worried that these men, for all their devout subscription to the Protestant faith, were undermining Christianity with their hypotheses, for instance, about the great age of the earth and the nonexistence of a universal flood. The Darwin of the late 1830s—note the date, for this will come up again in a moment—was a teleologist through and through: the organic world was inherently purposeful.

Natural selection spoke directly to this concern. Darwin was not so much interested in the overall purpose of an animal. That would get him into the kinds of ends that he thought the machine model barred. He was very interested in the slave-making propensities of the ants and saw the slaves as serving the ends of the masters. The slaves did not have their qualities in order to serve the masters. These qualities were appropriated by the masters to their ends. Of course, organisms survive and reproduce—the lucky or fitter ones do at least—but they don't do so with some overall purpose, at least not within the confines of science. It is different when we turn to the individual. The parts of organisms serve the ends of the whole organism. Eyes are for seeing because organisms with eyes are (in the appropriate circumstances) a lot better off than organisms without eyes. Note the qualification, for Darwin never thought that the needs of organisms are always the same. Mammals that live in caves might be blind—a good thing so the membranes don't get infected—whereas close relatives living above ground might have acute eyesight. Overall, final causes matter, and natural selection speaks to this. "Why, if man can by patience select variations most useful to himself, should nature fail in selecting variations useful, under changing conditions of life, to her living products? What limit can be put to this power, acting during long ages and rigidly scrutinising the whole constitution, structure, and habits of each creature,—favouring the good and rejecting the bad? I can see no limit to this power, in slowly and beautifully adapting each form to the most complex relations of life."[17]

Note that Darwin does not imply that end-directed characteristics—adaptations—are going to be perfect. He made that point very clear in later editions of the *Origin*. "Natural selection tends only to make each organic being as perfect as, or slightly more perfect than, the other inhabitants of the same country with which it comes into competition. And we see that this is the standard of perfection attained under nature. The endemic productions of New Zealand, for instance, are perfect one compared with another; but they are now rapidly yielding before the advancing legions of plants and animals introduced from Europe." He stresses that "natural selection will not produce absolute perfection, nor do we always meet, as far as we can judge, with this high standard under nature. The correction for the aberration of light is said by Müller not to be perfect even in that most perfect organ, the human eye."[18] Nor does Darwin think that everything has to be an adaptation. Homologies are not, for a start. For a second, the discussion of useless organs underlines the point. Sometimes things were of value but no longer. Sometimes things were never of value. With natural selection it is winning that counts. Not some overall perfection or utility. In the land of the blind, the man with one eye is king.

Relative, imperfect, or whatever, final cause counts. We spoke just above of the Darwin of the 1830s. This was the British heyday of final cause. By the 1850s the anatomists—Richard Owen and the newly arrived Thomas Henry Huxley (who may have quarreled with Owen but who shared much the same methodology)—were downplaying final cause of the traditional (utilitarian) kind and (obviously encouraging Whewell) making more and more of homology.[19] Darwin publishing in 1859 would have none of this. He had used homology throughout his long and detailed study of barnacles, but it was not where we find the all-important Newton-like causes.

> It is generally acknowledged that all organic beings have been formed on two great laws—Unity of Type, and the Conditions of Existence. By unity of type is meant that fundamental agreement in structure, which we see in organic beings of the same class, and which is quite independent of

their habits of life. On my theory, unity of type is explained by unity of descent. The expression of conditions of existence, so often insisted on by the illustrious Cuvier, is fully embraced by the principle of natural selection. For natural selection acts by either now adapting the varying parts of each being to its organic and inorganic conditions of life; or by having adapted them during long-past periods of time: the adaptations being aided in some cases by use and disuse, being slightly affected by the direct action of the external conditions of life, and being in all cases subjected to the several laws of growth. Hence, in fact, the law of the Conditions of Existence is the higher law; as it includes, through the inheritance of former adaptations, that of Unity of Type.[20]

Plato, Aristotle, Kant?

Think a little more about the nature of purpose in Darwin's theorizing at this level. It turns out to be a little more complex than you might expect and really quite interesting. We have cast our discussion in terms of three kinds of teleology or purpose. There is the Platonic kind, where the deity gets involved directly and does the designing. There is the Aristotelian kind, where there is a kind of internal force or spirit that directs things—perhaps more a principle of ordering. And then there is the Kantian heuristic kind, fully compatible with mechanism, where in some sense we impose the organization on the world, projecting ends because we cannot do biology without them. We might think that we should go straight to the Kantian position, and there is good reason to say this. Rather cleverly, Richard Dawkins called one of his books *The Blind Watchmaker*,[21] referring to the fact that Darwin took seriously Paley's claim that the organization of the living world is design-like but supplied a nonthinking mechanical explanation, namely, natural selection. In an earlier book, *The Selfish Gene*, Dawkins referred to organisms in a way that makes the mechanistic philosophy of post-Darwinian evolutionary biology quite explicit. "We are survival machines, but 'we' does not mean just people. It embraces all animals, plants, bacteria, and viruses."[22] This is all very much a position that has taken God and vital forces and those sorts of things out of the equation. Molecules in motion is all we have.

In the case of Darwin himself, we must be careful not to rush to conclusions and assume that he was so wise and prescient as to think at once like Richard Dawkins. Whatever his specific thinking about Plato—actually, as an educated Englishman of his day, he had a pretty good grasp of the central elements of the philosophy—the Anglican and then the deistic God loomed large in his thinking up to and including the publication of the *Origin*. Darwin thought that God had created the world and its contents, and this shows in his theorizing. In early versions of the theory (written in the early 1840s), God as creator comes through quite explicitly. "Who, seeing how plants vary in garden, what blind foolish man has done in a few years, will deny an all-seeing being in thousands of years could effect (if the Creator chose to do so), either by his own direct foresight or by intermediate means,—which will represent ⟨?⟩ the creator of this universe."[23] In the *Origin*, as many have noted, the Creator comes up again and again, and not referred to in a redundant or sarcastic way. "Authors of the highest eminence seem to be fully satisfied with the view that each species has been independently created. To my mind it accords better with what we know of the laws impressed on matter by the Creator, that the production and extinction of the past and present inhabitants of the world should have been due to secondary causes, like those determining the birth and death of the individual."[24]

However, all the time, as this last quotation shows, Darwin wanted to make the Creator work at a distance, through unbroken law rather than through a continuous set of miraculous interventions. Writing to his good friend, the American botanist Asa Gray, just after the *Origin*, Darwin said, "I am inclined to look at everything as resulting from designed laws, with the details, whether good or bad, left to the working out of what we may call chance." Elaborating: "I can see no reason, why a man, or other animal, may not have been aboriginally produced by . . . laws; & that all these laws may have been expressly designed by an omniscient Creator, who foresaw every future event & consequence."[25] It almost seems as though Darwin wants a Platonic overall causal picture, but one where his God has created a kind of Aristotelian world that has its own built-in purposes and can now be left to get on with the job. Parenthetically, late in life, Darwin spoke warmly of Aristotle, but although this was very sincerely meant, I am not sure how much direct influence we should seek from it.[26]

His was not really a stable position, and in any case, increasingly, Darwin gave up on any kind of God. This happened less because of the science and more for the kinds of reasons detailed in chapter 4. Like many Victorians, Darwin found Augustine's legacy morally repulsive. His father was the best man he had ever known. Was he to be condemned to everlasting hellfire for not being a believer? As God went, so the science got ever more secular. In a later edition of the *Origin*, Darwin wrote, "It has been said that I speak of natural selection as an active power or Deity; but who objects to an author speaking of the attraction of gravity as ruling the movements of the planets? Every one knows what is meant and is implied by such metaphorical expressions; and they are almost necessary for brevity." He stressed that "I mean by Nature, only the aggregate action and product of many natural laws, and by laws the sequence of events as ascertained by us."[27] Within a few years of the *Origin*, Darwin was pushing toward a strong Kantian kind of position, where all of the talk of ends and purposes and final causes is heuristic, something imposed from without rather than something discovered within. Remember, in the end this is not a man who is bothered about religion. He is more interested in getting on with his science. After he published the *Origin*, Darwin did a study of orchids.[28]

That is the kind of science he enjoyed, where he could get stuck into a group of organisms and the nitty-gritty of how they function. God, intentions, forces—divine, occult, vitalistic, whatever—just don't figure.

Although note—Darwin was ahead of Kant. The philosopher pushed God out of science, but he still didn't know why biology demanded final-cause explanation. Or rather, he did: it was because of God, but Kant couldn't bring this into his science, so he turned rather unpleasantly on biology itself. Darwin was never an atheist; at most an agnostic. However, to refer yet one more time to Richard Dawkins, Darwin made it possible to be an "intellectually fulfilled atheist."[29] Darwin gave a scientific explanation of final cause—of the purposeful nature of organic characteristics—without reference to God (or to Aristotelian self-organizing forces) and without having to suppose that God (or such forces) were hovering unseen in the background.

Darwin on Biological Progress

Darwin was deeply committed to the cultural ideology of progress and to the belief in biological progress, something that ends not just with human beings but with Europeans, preferably English capitalists. This thinking comes through again and again, most forcibly in the final words of the *Origin*.

> Thus, from the war of nature, from famine and death, the most exalted object which we are capable of conceiving, namely, the production of the higher animals, directly follows. There is grandeur in this view of life, with its several powers, having been originally breathed into a few forms or into one; and that, whilst this planet has gone cycling on according to the fixed law of gravity, from so simple a beginning endless forms most beautiful and most wonderful have been, and are being, evolved.[30]

The one thing of which we can be stone-cold certain is that, from the first, Darwin did not take biological progress as an easy given. "It is absurd to talk of one animal being higher than another.—We consider those, when the intellectual faculties[/] cerebral structure most developed, as highest.—A bee doubtless would when the instincts were—" It really didn't seem that humans had an exclusive lien on the summit. "People often talk of the wonderful event of intellectual man appearing.—the appearance of insects with other senses is more wonderful; its mind more different probably, & introduction of man nothing compared to the first thinking being, although hard to draw line.—" Above all, you have to work for success. It is not guaranteed. "The enormous number of animals in the world depends on their varied structure & complexity.—hence as the forms became complicated, they opened fresh means of adding to their complexity.—but yet there is no necessary tendency in the simple animals to become complicated although all perhaps will have done so from the new relations caused by the advancing complexity of others."[31]

Although Darwin sounds a bit iffy about calling something "higher" or "lower"—he repeated this warning on the flyleaf of Chambers's *Vestiges* when he read it—he obviously thought in terms of higher and lower because he used this language in the final paragraph of the *Origin*. Complexity is important but not any kind of complexity; more a kind of differentiation and specialization. The key here is Adam Smith's division of labor—also a notion of the French biologist Henri Milne Edwards. The higher you go, the more the parts have their own functions. "Von Baer's standard seems the most widely applicable and the best, namely, the amount of differentiation of the different parts (in the adult state, as I should be inclined to add) and their specialisation for different functions; or, as Milne Edwards would express it, the completeness of the division of physiological labour."[32] But can selection bring this about? Darwin thought it could through a process that today we speak of as "arms races." Lines of organisms compete against each other and their adaptations get ever more sophisticated.

> If we look at the differentiation and specialisation of the several organs of each being when adult (and this will include the advancement of the brain for intellectual purposes) as the best standard of highness of organisation, natural selection clearly leads towards highness; for all physiologists admit that the specialisation of organs, inasmuch as they perform in this state their functions better, is an advantage to each being; and hence the accumulation of variations tending towards specialisation is within the scope of natural selection.[33]

We can leave until chapter 6 whether this kind of solution will work. Important for us now is to see that even if, at times, one feels a little as if Darwin is like Moses—he led his children to the Promised Land but never got there himself—he knew the direction in which he was headed. He wanted a purely naturalistic understanding of purpose—of final cause, of teleology—and he provided the tools to get this. Natural selection explains purpose at the individual level. Natural selection–fueled arms races explain purpose at the historical level. Where does this leave us today? Let us see.

CHAPTER SIX

Darwinism

CHARLES DARWIN published *On the Origin of Species* in 1859 and *The Descent of Man* in 1871. He changed the world. Although there were those who continued to stand firm against evolution—indeed, as is well known, there are still those who continue to stand firm against evolution—generally, even the religious accepted that organisms, including humans, are the end point of a long, slow process of natural development. As in the Hans Christian Andersen tale about the lad who said openly that the king has no clothes, so when Darwin said "evolution," nigh everyone said that they had known it all along! Natural selection had more mixed success. Everyone accepted it to some extent. Huxley, for instance, always had some doubts about its universal power and applicability, but when it came to humans physically, he was fully convinced of its overwhelming importance. This said, the scientific community was slower in coming to full acceptance, and it was more in the popular domain that natural selection—and even more sexual selection—was a huge success. Poets, novelists, politicians, and many others harped on and on about its importance. Thus, the poet Constance Naden (1858–89), joking about these things—poking fun at young people of both sexes—in a delightful burst of mock despair wrote:

I HAD found out a gift for my fair,
    I had found where the cave men were laid:
Skulls, femur and pelvis were there,
    And spears that of silex they made.

But he ne'er could be true, she averred,
    Who would dig up an ancestor's grave—
And I loved her the more when I heard
    Such foolish regard for the cave.

My shelves they are furnished with stones,
    All sorted and labelled with care;
And a splendid collection of bones,
    Each one of them ancient and rare;

One would think she might like to retire
    To my study—she calls it a "hole"!
Not a fossil I heard her admire
    But I begged it, or borrowed, or stole.

But there comes an idealess lad,
    With a strut and a stare and a smirk;
And I watch, scientific, though sad,
    The Law of Selection at work.

Of Science he had not a trace,
    He seeks not the How and the Why,
But he sings with an amateur's grace,
    And he dances much better than I.

And we know the more dandified males
    By dance and by song win their wives—
'Tis a law that with avis prevails,
    And ever in *Homo* survives.

Shall I rage as they whirl in the valse?
    Shall I sneer as they carol and coo?
Ah no! for since Chloe is false
    I'm certain that Darwin is true.[1]

Modern Evolutionary Biology

From pseudoscience to popular science. When was evolutionary theory to become professional science, in the sense of something studied in university departments and with senior researchers and graduate students, grants, journals, and so forth? This happened starting around 1930 and picked up—particularly in England (where it became known as neo-Darwinism) and in America (where it became known as the synthetic theory of evolution)—over the next decades.[2] By 1959, somewhat arbitrarily choosing the hundredth anniversary of the *Origin*, one had (to use a somewhat hackneyed term) a fully functioning paradigm.

This was a Darwinian theory, in the sense that natural selection played (and continues to play) the central causal role, a status brought about by the melding of selection with the newly found and developed theory of heredity, Mendelian (and then later molecular) genetics. At the beginning of the twentieth century, the work of the somewhat obscure Moravian monk Gregor Mendel was rediscovered, and with this, the big hole in Darwin's theorizing could be filled.[3] Thanks particularly to the work in the second decade of the century by Thomas Hunt Morgan and his associates at Columbia University, it was seen that the crucial unit of heredity—the gene—is a physical thing (now known to be long threads of nucleic acid) on the chromosomes in the nuclei of cells. These genes maintain their integrity from generation to generation, thus giving selection something stable and heritable on which to act. However, every now and then the genes spontaneously change ("mutate")—much is now known about the causes but the important thing is that Darwin's insight was correct, the changes are random both in not appearing to order and in not necessarily bringing on new features of any use to the possessor.

Adaptations—characteristics with ends, with purposes—are as vital to modern evolutionary biology as they were to Darwin. Final-cause talk, thinking of organisms in terms of design, is all-important. One thing realized by today's evolutionists is that

Darwin was unduly pessimistic in thinking (as he did) that we will never see natural selection in action. In the right circumstances, it is readily observable. A justly celebrated demonstration of selection in action producing features that are directed toward ends is that of the couple Peter and Rosemary Grant and their long-term study of *Geospiza*—better known as "Darwin's Finches"—on an islet in the Galapagos Archipelago.[4] The Grants demonstrated unambiguously that in times of plenty, the beaks of the species they were studying were relatively fine and all-purpose—for cracking seeds, eating insects, and whatever—but that during times of drought, when the only available foodstuffs seemed to be hard-shelled nuts, the beaks evolved in a direction of stubbiness and strength. To put the matter teleologically: Why did the successful finches have stubby beaks? For the purpose of breaking up nuts with hard shells. And that was a very good thing from the viewpoint of the finches.

Paleontologists are into this game too. They think in terms of design, as if someone had sat down and built an organism to achieve a certain end. They are looking for purposes, for functions, for ends. A nice case in point is that of the strange noses of the duck-billed dinosaurs (hadrosaurs).[5] Flourishing some seventy-five million years ago, these were very peculiar-looking animals, with duck-like bills (very efficient for eating vegetation) and often very fancy crests on the skulls. In one group in particular (lambeosaurines), these crests were long, hollow growths, starting with the nose and going back across the head and sticking out at the back. After toying with a number of hypotheses—Could they be snorkels for foraging in water? Not likely, because they were essentially land animals—researchers narrowed their gaze to sexual selection and hypothesized that they were for producing noise to attract females. It seems that the structure would be ideal for this, and, in fact, the brutes reminded people of a trombone-like music maker, a medieval German wind instrument called the krummhorn. One can work out the physics of the airflow through the nasal tubes, and it turns out that the dinosaurs could produce a huge amount of noise, particularly at low frequencies. Honking hadrosaurs, to use a phrase. One should add that there are other bits and pieces of evidence supporting this hypothesis; for instance, we know a lot about their hearing apparatus (thanks to discovered ear bones), and all fits together very nicely.[6]

Refinements

It is worth noting two more points about modern thinking—extensions on Darwin's day. First, many worry (understandably) about how selection can possibly be effective if new variations are random in not arriving in time to order, as it were. A major advance in our thinking in this respect is due to the Russian-born American geneticist Theodosius Dobzhansky (1900–1975). He made much of what is known as "balanced superior heterozygote fitness."[7] Genes are paired with mates on corresponding chromosomes. Sometimes these genes are identical (homozygotes) and sometimes different (heterozygotes). An interesting situation ensues when heterozygotes do better in the struggle (are fitter) than either homozygote. A famous case in point concerns the awful genetic disease sickle-cell anemia. A person born a homozygote for a certain gene is going to die (without drastic medical intervention) from anemia, at the age of four. However, a person who is a heterozygote with one sickle-cell gene and one normal gene is going to be fitter than a homozygote for the normal gene. The reason is simple: namely, that being a heterozygote gives you a natural immunity to malaria, one not possessed by those with two normal genes. It is for this reason that the sickle-cell gene is found only in parts of Africa where malaria is endemic, or in populations from such areas, particularly North Americans of African descent. The important point to note is that no matter how bad the sickle-cell gene, it will persist in the population—in the "gene pool"—because of its virtues for heterozygotes.

Dobzhansky seized on this fact and generalized, assuming that such a phenomenon is widespread and that consequently any population is going to carry a large variety of genes—those from the same chromosome position are known as "alleles"—on which selection can act immediately without waiting for favorable mutations.[8] Superior heterozygote fitness is not the only putative way of getting in-group variation. Selection for rareness would also do the trick. Suppose a predator has to learn something about its prey before it can strike—color markings, for instance. A rare form would be at a selective advantage and thus start to spread, until it was so common that the predator would more quickly learn to seek it out, so positive selection would ease off and you might expect a balance between different forms. What is exciting is that at this point molecular biology (barely ten years after the discovery of the DNA model) came to the aid of organismic (evolutionary) biology by showing through new techniques (gel electrophoresis) that natural populations do harbor huge amounts of genetic variation.[9] Selection can indeed be effective. To use an analogy, imagine you were asked to write an essay on dictators for a course and the only source material available was the Book of the Month Club. You could wait for ten years for something, say on Hitler, to come up, by which time the deadline would have passed and you would have failed the course—or, analogously, gone extinct. But suppose you had a library at your disposal. If there was nothing suitable on Hitler, then perhaps there was something on Napoleon. Or on Stalin. Or on others. You might not be able to write on a topic you like, but you could write on something and pass the course. Similarly, if a new predator turns up, perhaps there is an adaptive-camouflage gene waiting in the gene pool. Or one that enables you to change ecological niches where you are now safe. Or something that makes you extremely unpalatable to the predator. There is no guarantee that you will not go extinct, but there is probably some useful tool in your tool box. All in all, selection is highly plausible as an important creative force.

The second point about modern evolutionary thinking is that no one (starting with Darwin, as we have seen) thinks that every last thing about organisms, living or dead, has to be (or had to be) adaptive. The late Stephen Jay Gould, paleontologist and popular science writer, made much of this. Greatly influenced by German biology, he focused on the homologies between organisms, stressing their importance for establishing the fact of evolution but their irrelevance for proving the force of natural selection.[10] More broadly, in a well-known article, coauthored by Richard Lewontin, "The Spandrels of San Marco," Gould argued strongly that Darwinian evolutionists assume far too readily that living nature is adaptive, that it is full of purpose.[11] He felt that evolutionists slide into some kind of panadaptationism, thinking that every last organic feature has to be functional, the product of natural selection. Referring to the Leibnizian philosopher in Voltaire's *Candide*, he accused evolutionists of Panglossianism, thinking that these must be the best of all possible features in the best of all possible worlds. And to make the case complete, supposedly, evolutionists invent "just so" stories—thus named from Rudyard Kipling's fantasy stories—with natural selection scenarios leading to adaptation.

As a counter, Gould (and Lewontin) drew attention to the triangular decorative aspects of the tops of pillars in medieval churches, arguing that although such "spandrels" seem adaptive, they are in fact by-products of the builders' methods of keeping the roof in place. "The design is so elaborate, harmonious, and purposeful that we are tempted to view it as the starting point of any analysis, as the cause in some sense of the surrounding architecture." This, however, is to get things precisely backward. "The system begins with an architectural constraint: the necessary four spandrels and their tapering triangular form. They provide a space in which the mosaicist worked; they set the quadripartite symmetry of the dome above."[12] Who knows but that we have a similar situation in the living world? Much that we think adaptive is merely a spandrel, and such things as constraints on development prevent anything like an optimally designed world. Perhaps things are much more random and haphazard—nonfunctional—than the Darwinian thinks possible.

A deal of ink—a very great deal of ink—was spilled over these claims. General reaction by Darwinian evolutionists—who make up perhaps 95 percent of this population—was that much that Gould said was true but well-known already.[13] One phenomenon bringing on the nonadaptive is so-called genetic drift. This is where the vagaries of breeding—the chance encounters between organisms—can be sufficiently powerful to counter the effects of selection. This was the basis of a theory put forward in the early 1930s by the American population geneticist Sewall Wright—the "shifting balance theory of evolution."[14] Drift is most likely to occur in small populations and, based on an extensive study of shorthorn cattle, he argued that evolution proceeds by drift, creating innovative new features when large populations are fragmented, with these features then spreading through the whole group when the fragmentation comes to an end. As it happens, there has been much criticism of Wright's overall theory.[15] Yet no one denies that drift probably does have some role—and it is agreed that it probably has a major role at the molecular level below the winnowing effects of the struggle for existence. (I will discuss this in more detail later.) More generally, no one denies that many features are going to be nonadaptive, or perhaps were once adaptive and no more. Why do vertebrates have four limbs rather than six like the insects? John Maynard Smith argued that this may be a relic of when vertebrates were aquatic and two limbs fore and two limbs aft were very effective for raising or lowering the body immersed in water.[16] There is nothing sacrosanct about numbers. There are some fossil vertebrates with eight or nine digits rather than five.

So where are we today in evolutionary thinking? Don't go away with the message that, whatever they may hope for of the Design argument, biologists today are now questioning seriously what was labeled the first part of the argument, to the design-like nature of the world. In the world of organisms, adaptation is the norm—the hugely well-justified null hypothesis—and it is your task to make the contrary case if you so wish. Purpose thinking rules, and it is cherished.[17] In a good Kantian sense, today's biologists use end-directed thinking and language when they are dealing with organisms. The mountains on the moon have no purpose. The internal workings of the hadrosaur have a full and genuine purpose, to make a lot of noise of a particular kind. Going back in time, the young hadrosaur is here and now. The noise and the sexual combat are in the future. The bodily structure is to be explained in terms of growth and physiology (efficient causes) and of getting mates (final causes). And this is true even if the hadrosaur dies an unrequited virgin. Is it anything more than heuristic? Not really, but "anything more" is hardly the best way to think of it. Because the hadrosaur is design-like, it is appropriate to use the metaphor of design. There is no implication that there is a designer any more than there is an implication that you are really thin and slimy when I accuse you of having wormed your way into my affections and trust for your own nefarious ends. The metaphor remains, and Kant was right—you simply cannot do this kind of biology without it. In its own way, recognizing this is as much a challenge to or qualification of a simple, Cartesian, post–Scientific Revolution, mechanistic view of the world as is quantum mechanics. That is no small thing.

What then of value? We have seen that one of the defining marks of purpose talk is that it is enmeshed with value commitments. If something is directed to achieving some end, then in some sense the end is being valued and that which is helping to achieve it has value because of its role. We want, we desire food, and teeth are a means of chewing and subsequently digesting that food. The food has value for us and the teeth are valuable inasmuch as they enable us to use that food. However, this does raise a serious problem or at least question. Are we not in some sense finding values in the world—or perhaps in a Kantian sense imputing values to the world? And this is not to go against the metaphysics of modern science, which fully endorses Hume's distinction between "is" and "ought." Ontologically, the world as such is *res extensa*, molecules in motion. It has no intrinsic value.

My fellow philosophers have spent many happy hours analyzing this problem.[18] Medieval theologians worrying about the number of angels who can dance on the head of a pin have nothing on analytic philosophers when they gear up—except, unlike today, the story of the scholastics is probably a calumny by later writers. Like me (as I mentioned in my acknowledgments), brought up in neo-Humean traditions, trying to avoid talk of values, philosophers have tied themselves in knots, meriting attention by Hilaire Belloc.

> The chief defect of Henry King,
> Was chewing little bits of string.
> At last he swallowed some that tied,
> Itself in ugly knots inside.

Alas, that was the end of poor Henry, not—one is relieved to say—the fate of the philosophers. From the viewpoint of finding adequate solutions, it might have been, for in trying to analyze purpose or function without reference to value, one is trying to square the circle. One popular neo-Kantian attempt lays itself open to all sorts of counterexamples—unable to distinguish between the heart pumping in order to circulate the blood, which does have value, from the heart pumping in order to make sounds, which does not have value.[19] Another popular attempt simply ignores the end focus of purpose statements, which may perhaps cure the disease but at the expense of the patient's life.[20]

If I sound cynically critical of others, the answer is (as so often) because I see the failings of my earlier self. I saw something was not right, but, too committed to my philosophical paradigm, the best I could suggest was one-up on ignoring the problem. I urged the dropping of all end talk.[21] Which is one way of solving the problem, I suppose. Not exactly a solution appealing to your average paleontologist, let alone game strategist or moral educator. Although the spirit of my earlier self continues. One recent attempt to analyze function focuses on organisms maintaining themselves and ends up by saying explicitly that the reproductive organisms, with respect to their possessors, have no functions.[22] Tell that to students in your freshman classes. Truly, the right move forward is not to deny or cover up the value component of teleological understanding but to embrace it fully. To be fair, I am not alone in now stressing the importance of values. One who has done so is the philosopher Mark Bedau, who notes explicitly that going back through the history of philosophy makes transparently clear this point about the necessity of a value-analysis. "It has ancient roots in Plato and Aristotle, and its modern exponents include Leibniz and Kant."[23]

Interestingly, but perhaps not surprisingly, those whom Bedau lists from today's thinkers who have sensed the need of a value-analysis have, like the practicing biologist and sometime Dominican priest Francisco Ayala,[24] a strong grasp of the history of philosophical thought. Obviously, if we are in a natural world, rather than in God's world, there has to be a change in how we are to approach the problem of value. But this can be seen and tackled. The right move here is to distinguish "value" from "evaluation." What the modern scientist denies is any kind of absolute or overall (externally conferred) value to the world—the scientist acting as scientist, that is. (Often this is referred to as an implication of "methodological naturalism," as opposed to "metaphysical naturalism," which would deny any such value under any perspective.) This does not mean that the modern scientist cannot make value judgments in a comparative sense. He or she might judge one kind of internal combustion engine as a great deal more efficient than another. That is obviously a value judgment—not absolute because you might judge that in this day and age, internal combustion engines are never a good thing—but comparative in the context. Pushing the argument, a biologist might argue that a group of organisms was able to take over from its rivals because it developed a mode of movement or of internal functioning a great deal more efficient than that of its competitor. This again is a value judgment—not absolute because the newly successful organism might do horrendous damage overall. Think of rabbits in Australia. It is a comparative judgment and in the circumstances fully legitimate.[25]

Conflict over Progress

Turn now to purpose in history. You might think that this is going to be a very short discussion. Natural selection is opportunistic. What works in one situation does not necessarily work in another. There is no reason to expect a forward direction to evolution, even one interrupted by reversals and sidestepping. Moreover, Mendelian/molecular genetics is adamant. There is no direction to the new variations—the building blocks—of evolution. No "higher" or "lower." That is an absolute. There is no value in the course of evolution. Stephen Jay Gould was eloquent. There is no direction and so evolution apparently can go whichever way. Progress to humans is just not on. "A noxious, culturally embedded, untestable, nonoperational, intractable idea that must be replaced if we wish to understand the patterns of history."[26] Making facetious reference to that celestial body that hit the earth sixty-six million years ago, wiping out the dinosaurs and making possible the Age of Mammals, Gould wrote: "Since dinosaurs were not moving toward markedly larger brains, and since such a prospect may lie outside the capabilities of reptilian design . . . we must assume that consciousness would not have evolved on our planet if a cosmic catastrophe had not claimed the dinosaurs as victims. In an entirely literal sense, we owe our existence, as large and reasoning mammals, to our lucky stars."[27]

Yet for all the sepulchral warnings one hears on the subject, thoughts of progress have a nasty way of creeping back in. My favorite was the American Museum of Natural History in New

York City a year or two back. Down in the basement was a display about human evolution with all sorts of careful caveats about not believing in biological progress. The floor above had the Hall of Mammals, going from the shrew at one end to the great apes at the other. The Muséum National d'Histoire Naturelle in the Jardin des Plantes on the left bank of the Seine in Paris is even less circumspect. The top floor has a display of human culture and technology, going progressively from the primitive to the present. The floor below has a happily progressive display of evolution, ending with the visitor, him- or herself, on the television screen on exiting. Who is about to deny progress under these circumstances?

Interestingly, Gould notwithstanding—and we shall see that he is more complex and convoluted than one might expect on first sight—many of today's leading evolutionists are quite open about their beliefs in biological progress. The most distinguished member of the fraternity, Edward O. Wilson, Harvard professor and world-leading specialist on ants and on sociobiology (the evolution of social behavior), is unequivocal. "The overall average across the history of life has moved from the simple and few to the more complex and numerous. During the past billion years, animals as a whole evolved upward in body size, feeding and defensive techniques, brain and behavioral complexity, social organization, and precision of environmental control—in each case farther from the nonliving state than their simpler antecedents did."[28] Adding: "Progress, then, is a property of the evolution of life as a whole by almost any conceivable intuitive standard, including the acquisition of goals and intentions in the behavior of animals." Elsewhere he writes of the "pinnacles" of social evolution, judging that we humans have won that competition outright.[29]

Part of the ongoing problem is that of defining biological progress in terms that are not flagrantly circular. If you define progress in terms of being humanlike, which is basically the move of Wilson, it is hardly surprising to find that we have won.

Often complexity is thought to be the key, but here there are difficulties. In many respects, for instance, humans are a lot less complex than was once supposed. For instance, some single-celled organisms have more DNA than we do! Apart from anything else, in biology, as in real life, "Keep It Simple Stupid" is often a very good motto. The evolutionary biologist George Williams, a lifelong opponent of biological progress,[30] was fond of pointing out that in principle the jet engine is far simpler than the internal combustion engine. And all of this is apart from the very difficulties in defining complexity. Always ready with an answer, Dawkins suggests that it is just a matter of description. If you write down all of the features of one organism and compare it with a like description of the features of another organism, the one with the longer list wins. "If you have a lobster and an earthworm and you wish to decide which is the more complex, proceed as follows. Write a book about the lobster, write another book about the earthworm, and count the number of words in the toolbox. The animal which needs the larger book is the more complex."[31] The trouble here is the same as what taxonomists ran into fifty years ago with the advent of computers—it all seemed so easy. Count up the characteristics, put them into the machine, and out would come objective classifications. Rapidly people saw that this approach—"phenetic" or "numerical" taxonomy—would not work because no one knew how to divide up the countable characteristics.[32] Does a bald man differ from one of the Beatles by one feature or by literally thousands—hair by hair? This isn't to say that the lobster is not more complex than the earthworm, but that it isn't easy to say on what basis.

In the *Origin*, Darwin admitted candidly that paleontologists have a sense of progress. "The inhabitants of each successive period in the world's history have beaten their predecessors in the race for life, and are, in so far, higher in the scale of nature; and this may account for that vague yet ill-defined sentiment, felt by many palæontologists, that organisation on the whole has progressed."[33] Despite Gould, many of today's paleontologists feel the same way. The late Jack Sepkoski, one of the most highly regarded paleontologists of the end of the last century—actually they preferred to call themselves "paleobiologists" to denote the fact that they wanted to move beyond simply digging out fossils and to understanding the past in (evolutionary) biological terms—put things in a very American way, referring to the opening of the frontier. He was much interested in mass extinctions, which he saw as both creative and destructive.

> Mass extinctions have probably been good for the evolving biosphere. I said, "good" and I've got to explain why I said "good"—in the sense that they probably promoted diversity.
>
> Real evolutionary innovations, probably coming in during the rebound of these extinction events, clear out a lot of diversity. Clear out a lot of biomass. We're back into semi-frontier days. Sort of environment where you don't have to be real good to get on, so something very new and different may be able to grab hold of a piece of the ecological pie, and hold it, giving rise to new kinds of organisms.
>
> So mass extinctions are good in that sense. They promoted evolutionary innovation.[34]

Generally, when talking of change, paleobiologists slip into value-impregnated language and progress-type talk. For instance, a fairly typical discussion of evolution runs: "Fish can be seen to have undergone significant morphological 'advancement' [for example from the chondrostean to holostean grade among Actinopterygii, or the 'cladodont' to 'hybodont' grade in Chondrichthyes]."[35] Analogously, in the plant world we find distinction between primitive and advanced, with the former characterized by the leaves exhibiting " 'first rank' leaf architecture: i.e., poor definition of vein orders, irregularity of spacing, angle of departure, course, and branding patterns of secondary and higher-order veins, and incomplete differentiation of blade and petiole, a syndrome of characters originally postulated to be primitive in dicots on the basis of comparative studies of Recent forms."[36]

What is striking is how most evolutionists more or less take biological progress for granted—for all that they are prone to deny it when in public and totally sober—and go on to argue from there. There are a number of reasons for this. One is that since we are asking questions about progress, necessarily we are at the end of the evolutionary process and so there is an inclination to think that we must have won. There tends not to be much fellow feeling with warthogs at a time like this. Second, bound up with this first point is the sense that if warthogs feel that they are so very important, why don't they speak up and say so? Because we can ask questions about progress, we tend to judge progress in these terms. Perhaps warthogs judge progress in terms of wallowing in mud and letting the world pass by except when out feeding or copulating. Would a warthogian Aristotle judge that inferior? Third, in a tradition going back to Diderot, there is a tendency to read hopes of cultural progress into the biological world and come up with confirmatory biological progress. Uniquely, scientists live in a world where—the social constructivists and other relativists notwithstanding—there is real progress. Newton was better than Aristotle and Einstein was better than Newton. Creationism is wrong and evolution is right. Why wouldn't one expect to find biological progress, especially since we are the ones responsible for scientific progress?

Arms Races

Yearning sentiments aside, is there any reason, any Darwinian reason, to think that progress will occur? Specifically, is there any Darwinian reason to think that biological processes will lead to human beings? Or, at least, is there any Darwinian reason to think that biological processes will lead to what have been called "humanoids," that is, humanlike beings with intelligence and so forth? I don't suppose anyone is demanding that we have five rather than six digits or white/brown/black skin rather than blue or green skin. I am not so sure about sex, in part because no one has a fixed idea about the causes of sex. It certainly seems to be the case, however, that sex makes things happen, namely, gathering good mutations together quickly in one individual, and so one suspects it unlikely that our humanoids would be sexless entirely. But there could be variations, like them all being hermaphrodites, all both capable of fertilizing and open to being fertilized and giving birth.

Humans have evolved, so obviously they could evolve. If you were to allow the hypothesis of multiverses—an infinite number of other universes, some (presumably an infinite number) like ours—then presumably (updating the argument of the atomists) somewhere, sometime, humans were going to evolve. To say otherwise is to say that they couldn't. In fact, one presumes that an infinite number of humans are going to evolve. I remember once the late J.J.C. ("Jack") Smart, a British-born, absolute fanatic about cricket—and a man who, incidentally, educated literally thousands of young Australians—saying with some glee that all over the universe there is an infinite number of teams capable of beating the Australians! This, one should say, was said at a time when it seemed that nothing natural was ever going to beat the Australians. Smart, incidentally, felt even more strongly on the subject of cricket played in colors other than the traditional white. Presumably, all over the universe there are teams clad in shocking-pink hot pants, capable of beating the English by an innings and several wickets.

However, even if this is so—and there are serious critics of multiverses—we hardly have anything one would be inclined to call "progress." It starts to sound like huge arrogance to say that, in a situation where presumably one has billions of life-forms, everything in some sense bows down to us—or even to the billions of us in various galaxies. We can put the values in, if we want, but we are not reading them out of nature. There is nothing in biology itself to say we are better. Although, this said, there is equally nothing to stop us looking for Darwinian reasons for thinking that the nature of the evolutionary process is such that humanlike beings are (best scenario) necessarily going to emerge or (less attractive scenario) at least very likely to emerge. There are two popular proposals.

The first builds on the insight of Darwin about competition leading to improvement, particularly the competition between evolving lines leading to improvement. This idea about "arms races" was elaborated in most detail by Julian Huxley in a little book at the beginning of the last century. He gave a graphic description of an arms race couched in terms of the then state-of-the-art naval military technology. "The leaden plum-puddings were not unfairly matched against the wooden walls of Nelson's day."[37] Now, however, obviously having in mind the then huge naval competition between Britain and Germany, "though our guns can hurl a third of a ton of sharp-nosed steel with dynamite entrails for a dozen miles, yet they are confronted with twelve-inch armor of backed and hardened steel, water-tight compartments, and targets moving thirty miles an hour. Each advance in attack has brought forth, as if by magic, a corresponding advance in defence." Likewise in nature, "if one species happens to vary in the direction of greater independence, the inter-related equilibrium is upset, and cannot be restored until a number of competing species have either given way to the increased pressure and become extinct, or else have answered pressure with pressure, and kept the first species in its place by themselves too discovering means of adding to their independence." Eventually: "it comes to pass that the continuous change which is passing that through the organic world appears as a succession of phases of equilibrium, each one on a higher average plane of independence than the one before, and each inevitably calling up and giving place to one still higher."

One who has (without acknowledgment) picked up enthusiastically on this kind of thinking is Richard Dawkins. "Directionalist common sense surely wins on the very long time scale: once there was only blue-green slime and now there are sharp-eyed metazoan."[38] He too finds the key in arms races. As one who embraced computer technology early and enthusiastically, perhaps expectedly Dawkins notes that, more and more, today's arms races rely on computer technology rather than brute power, and—in the animal world—he finds this translated into ever-bigger and more efficient brains. No need to hold your breath about who has won. Dawkins invokes a notion known as an animal's EQ, standing for "encephalization quotient."[39] This is a kind of cross-species measure of IQ that takes into account the amount of brain power needed simply to get an organism to function (whales require much bigger brains than shrews because they need more computing power to get their bigger bodies to function), and that then scales according to the surplus left over. Dawkins writes, "The fact that humans have an EQ of 7 and hippos an EQ of 0.3 may not literally mean that humans are 23 times as clever as hippos! But the EQ as measured is probably telling us *something* about how much 'computing power' an animal probably has in its head, over and above the irreducible amount of computing power needed for the routine running of its large or small body."[40]

As always, it is the analogy with human progress that is the key.

> Computer evolution in human technology is enormously rapid and unmistakably progressive. It comes about through at least partly a kind of hardware/software coevolution. Advances in hardware are in step with advances in software. There is also software/software coevolution. Advances in software made possible not only improvements in short-term computational efficiency—although they certainly do that—they also make possible further advances in the evolution of the software. So the first point is just the sheer adaptedness

> of the advances of software make for efficient computing. The second point is the progressive thing. The advances of software, open the door—again, I wouldn't mind using the word "floodgates" in some instances—open the floodgates to further advances in software.[41]

He adds, "I was trying to suggest, by my analogy of software/software coevolution, in brain evolution that these may have been advances that will come under the heading of the evolution of evolvability in the evolution of intelligence."[42]

Others endorse similar lines of thinking. For instance, there is reason to think that shellfish are in arms races with predators, putting ever-greater resources into thicker, tougher shells, with the predators developing ever-more efficient methods of boring into shells and extracting the contents. However, not every Darwinian biologist is that enthused by arms races. The fossil evidence, for instance, does not show unambiguously that prey and predators have become ever faster. And even if arms races are ubiquitous, it does not follow that intelligence will always emerge. Having high intelligence means having large brains, and having large brains means having ready access to large chunks of protein, the bodies of other animals. There were no vegans in the Pleistocene. Sometimes—as cows and horses demonstrate—it is just easier to get your food in other ways, especially if you are living on grassy savannahs. Despite his enthusiasm for progress, Jack Sepkoski put matters colorfully and definitively: "I see intelligence as just one of a variety of adaptations among tetrapods for survival. Running fast in a herd while being as dumb as shit, I think, is a very good adaptation for survival."[43] So the overall answer seems to be that although arms races may well lead to intelligence, there is no guarantee that this will happen, and given that we have only the one instance (admittedly successful) to go on, it would be rash to argue with too much confidence that progress up to humans is the norm.

Channeling

The other favored approach to getting progress out of the Darwinian system works on the theme of ecological niches and organisms finding them and occupying them. We often think of the broad niches occupied by organisms—animals particularly—water, earth, and air. Why not just add on another—culture—and suppose that it was waiting to be occupied and finally protohumans found it and moved in? Gould of all people floated some idea like this. He thought that, if not on our earth, then somewhere in the universe this might have happened. He quoted Theodosius Dobzhansky: "Granting that the possibility of obtaining a man-like creature is vanishingly small even given an astronomical number of attempts . . . there is still some small possibility that another intelligent species has arisen, one that is capable of achieving a technological civilization."[44] Gould commented, "I am not convinced that the possibility is so small." He gave an argument that evolutionary convergence (where two different lines evolve essentially similar adaptations to survive and reproduce) suggests that even though major intelligence has arisen but once on this earth, it is quite possible that elsewhere in the universe it has arisen quite independently. "But does intelligence lie within the class of phenomena too complex and historically conditioned for repetition? I do not think that its uniqueness on earth specifies such a conclusion. Perhaps, in another form on another world, intelligence would be as easy to evolve as flight on ours."[45]

I am not sure that one would want to use the word "progress" here. As in the multiverse discussion, that intelligence appears over and over does not really justify one in saying it is the best or even better than other life-forms. At the least, one wants some kind of channeling or funneling toward humans. Using much the same argument as Gould, the paleontologist Simon Conway Morris (as a Christian) is very keen to argue for the inevitability of the appearance of humans. He argues that only certain areas of what we might call "morphological space" are welcoming to life-forms (the center of the sun would not be, for instance), and that this constrains the course of evolution.[46] Again and again, as Gould argues, organisms take the same route into a preexisting niche. The saber-toothed, tigerlike organisms are a nice example, where the North American placental mammals (real cats) were matched right down the line by South American marsupials (thylacosmilids). There existed a niche for organisms that were predators, with catlike abilities and shearing/stabbing-like weapons. Darwinian selection found more than one way to enter it—from the placental side and from the marsupial side. It was not a question of beating out others but of finding pathways that others had not found.

Conway Morris argues that, given the ubiquity of convergence, we must allow that the historical course of nature is not random but strongly selection-constrained along certain pathways and to certain destinations. Most particularly, some kind of intelligent being was bound to emerge. After all, our very own existence shows that a kind of cultural adaptive niche exists—a niche that prizes intelligence and social abilities. "If brains can get big independently and provide a neural machine capable of handling a highly complex environment, then perhaps there are other parallels, other convergences that drive some groups towards complexity." Continuing: "We may be unique, but paradoxically those properties that define our uniqueness can still be inherent in the evolutionary process. In other words, if we humans had not evolved then something more-or-less identical would have emerged sooner or later."[47]

Does this do the trick and is this progress? One might question positive answers to both questions. Even if it exists, why should we or anyone else necessarily or even probably enter the culture niche? Life is full of missed opportunities. Maybe Gould is right and most times evolution would have gone other ways and avoided culture entirely. Warthogs rule supreme. Huxley always argued that now humans occupy the culture niche, no other animal is going to be able to enter.[48] Perhaps other animals (dinosaurs) would have prevented our animals (mammals and then primates) from making their way to the door. In any case, many wonder if it is right to think that niches are just waiting out there, ready to be conquered and entered. Do not organisms create niches as much as find them? There was hardly a niche for head lice, for instance, until vertebrates like us humans came along. Should we expect that there was a niche for culture, just waiting there, like dry land or the open air? Perhaps there are other niches not yet invented. We cannot imagine something other than consciousness; but take heed of the wise warning of J.B.S. Haldane: "Now my own suspicion is that the Universe is not only queerer than we suppose, but queerer than we *can* suppose."[49] For all their talk about analogy, Christians tend to think that their God can get up to some pretty clever tricks, way beyond their ken. Perhaps these are not all supernatural abilities, but simply abilities that were omitted from our evolution. Perhaps, far from being the best, we are a short side-path and very limited in the true scheme of things. No more than in the case of arms races do we get much guarantee of either human emergence or a sense that we are in some way superior and for this reason we won.

Purpose is there in Darwinian biology, through and through. Thanks to Darwin, many enthusiasts think we have come a long, long way. We have purpose in the individual feature—the eye exists in order to see. Equally, although there are some (including myself) who are not so enthusiastic on this score, many think we have purpose in history. Humans are the destined end point, thus far. Have we arrived at the bright, Elysian shore? Many Darwinians think we have. Others—who have greater or less degrees of enthusiasm for natural selection—are not so certain, as we shall now see.

CHAPTER SEVEN

Plato Redivivus

OLD HABITS DIE HARD. The same can be said of old philosophies. We have traced what many (I suspect most Darwinians) would regard as the triumph of the Kantian perspective. Muted perhaps, but triumphant nevertheless. An undeserved victory in the eyes and hearts of vigorous opponents. From the time of the Scientific Revolution to the present, we find vocal representatives of what I am characterizing as the Platonic (external) approach or tradition and of the Aristotelian (internal) approach or tradition. Let us take them in turn.

Creator God

Before the *Origin*, there were those like Whewell and Adam Sedgwick, professor of geology at Cambridge, who simply put down the origins of new species to divine intervention.[1] The fossil record shows that there has been a turnover of forms, and extinction is almost certainly due to natural causes. But when it comes to new forms, God intervenes miraculously. After the *Origin*, there were those who felt the same way. Louis Agassiz, Swiss-born ichthyologist and professor at Harvard, could never accept evolution, even though his students (including his son) stepped over the line pretty sharpishly.[2] The preferred option though, for those who were Christians believing in a Creator God, was some form of guided evolution. God puts direction into new variations and hence natural selection has at most a kind of garbage disposal function—it gets rid of the bad forms but does little or nothing to create new, good forms. This was the stance of the evangelical Presbyterian Asa Gray.

> But there is room only for the general declaration that we cannot think the Cosmos a series which began with chaos and ends with mind, or of which mind is a result: that, if, by the successive origination of species and organs through natural agencies, the author means a series of events which succeed each other irrespective of a continued directing intelligence—events which mind does not order and shape to destined ends—then he has not established that doctrine, nor advanced toward its establishment, but has accumulated improbabilities beyond all belief.[3]

Darwin would have nothing of this. Picking up on a metaphor used by Gray about water being channeled down certain streams, he wrote, "If we assume that each particular variation was from the beginning of all time preordained, the plasticity of organisation, which leads to many injurious deviations of structure, as well as that redundant power of reproduction which inevitably leads to a struggle for existence, and, as a consequence, to the natural selection or survival of the fittest, must appear to us superfluous laws of nature." Darwin spotted what was going on: "On the other hand, an omnipotent and omniscient Creator ordains everything and foresees everything."[4] Basically, although they went on arguing, this was it. Gray suggested that if you are building a house, you have to have the stones precut to order. Darwin countered that if you had enough, as drystone wall builders show us, you can do a very good job on what nature hands you. And so on and so forth.

As noted, another who sought divine direction in evolution was Wallace, although he appealed not to the God of the Christians but to some kind of World Spirit—"a superior intelligence has guided the development of man in a definite direction, and for a special purpose, just as man guides the development of many animal and vegetable forms." Continuing: "We must therefore admit the possibility that, if we are not the highest intelligences in the universe, some higher intelligence may have directed the process by which the human race was developed, by means of more subtle agencies than we are acquainted with."[5] Darwin, as we know, was not convinced, although he did agree that Wallace had made some good points, showing, for instance, how unlikely it is that human hairlessness and human intelligence came unaided through natural selection. This was the reason for pumping up sexual selection. Neither man budged from then on.

Being (what I am calling) a Platonist about these sorts of things did not mean one had to be a biblical literalist—six days of creation, six thousand years, worldwide flood, and that sort of thing. As it happens, given the science of his day, Augustine thought these reasonable beliefs—"we compute from the sacred writings that six thousand years have not yet passed since the creation of man"[6]—but he always insisted that advances in empirical understanding might mean modification of literal readings of scripture. As he pointed out, the ancient Jews could not be expected to understand the theories and experience of sophisticated Roman citizens. This was always the traditional Christian perspective. Wallace was an oddity, but no one could doubt the intensity or authenticity of Gray's Christian commitment. However, as is well known, not all of his fellow countrymen felt this way, and before and after the Civil War in the South, in particular, a form of evangelical Protestant literalism took root and throve. Not the least of the attractions of such a stance was that it was taken (with reason) to offer a biblical justification of slavery. After the war, the story of the Israelites in captivity was much appreciated. God heaps burdens most on those whom he loves most.[7]

No one was a total literalist. The whore of Babylon was rarely taken to be a historical figure and generally interpreted as the pope or the Catholic religion or some such thing. Also, biblical claims about a thousand years being but a day in the eye of the Lord were generally taken as reason to suppose an old earth. This changed after the Second World War, thanks on the one hand to the influence of Seventh-day Adventist theology—it always took the days as literal days and the earth's time span as 6,000 years—and on the other hand to the receptive nature of evangelical culture.[8] This was the time of the Cold War with fears of nuclear conflagration. Armageddon started to loom large in many minds, and this was reason for a "dispensationalist" theology—periods of time brought to violent ends, the first being the expulsion from Eden and the last and future being the End of Times. The Noachian Deluge taken literally was a key piece of evidence, and thanks to the enterprising authors of *The Genesis Flood* (1961)—John Whitcomb, a biblical scholar, and Henry Morris, a hydraulic engineer—a "Young Earth" literalism was promoted with much success.[9] Creation science, so-called because the insistence was that it was scientific and hence could circumvent the First Amendment–based prohibition on teaching Genesis in state-supported biology classes, owed little historically to Plato—who I am sure would be absolutely horrified that I am including it in a tradition started by him—but it made direct design, purpose, absolutely central. Design in the world of organisms, and then in the scriptures, put all in context and pointed the way forward for the believer. "We must go to the Scriptures for salvation. The scientific evidence for design and creation and the Creator are vital to present to those who do not know or believe the Bible (note Acts 14:15–17 and 17:22–29), but then they must go to the Scriptures if they would learn about the true God and His work of creation and redemption."[10]

Intelligent Design Theory

Creation science met its own Armageddon in the state of Arkansas in 1981, when—thanks to the testimony of a cohort of experts, including Francisco Ayala and Stephen Jay Gould—a federal judge ruled that it is religion, not science, and hence could not be taught in state-funded schools.[11] But like a phoenix from the ashes, Creationism Lite arose, happily named "intelligent design theory" (IDT), the brainchild of Harvard-educated, Berkeley law professor Phillip Johnson.[12] For some of its enthusiasts, IDT is a smoother version of Young Earth creationism. For others, it is the foundation of a form of "guided evolution" that Gray would have understood and appreciated—an evolution where God steps in continuously and keeps things on track. No matter, the point is that at its heart is purpose, and this is something imposed from without. Central to IDT is the claim that the organic world is so "irreducibly complex" that blind law could not in principle explain it. We must invoke a designer of some sort. Usually IDT supporters rush to say that they do not imply that the designer must be a Designer—a god or God, specifically the Christian God. In fact, they are being a bit disingenuous here, for by and large they do mean the Christian God. Of one thing you can be certain, they don't think the Designer is a grad student on Andromeda, fooling around with the life-forms, using Planet Earth for its dissertation.

Michael Behe, the best-known scientific supporter of IDT, makes much of bacteria, specifically, those that use a flagellum (a kind of whiplike strand) powered by a sort of rotary motor, to propel themselves along. Everything is highly complex and nothing will function until and unless every part is absolutely in its place. For example, the "flagellin," the external filament of the flagellum, is a single protein. It makes a kind of paddle surface that contacts the liquid during swimming. Near the surface of the cell, one finds a thickening—just as needed—so that the filament can be connected to the rotor drive. In turn, we need a connector, something known as a "hook protein." The filament has no motor. It has to be somewhere else. "Experiments have demonstrated that it is located at the base of the flagellum, where electron microscopy shows several ring structures occur."[13] From all of this, Behe concludes that the whole system is far too complex to have come into being in a gradual fashion. It had to be formed in one step, and such a process must involve some sort of designing cause. A similar argument is used of other phenomena, for instance, the blood-clotting cascade, a rather complicated sequential chemical process that takes place when you cut yourself and the gushing blood shortly starts to coagulate and stops pouring out. In both cases Behe claims there is no way that blind law could have created them. He argues this way primarily on the grounds that if all parts are not in place all at once, things do not work. He also, by illustration, uses the analogy of a five-part mousetrap, arguing that it too would not work unless all five parts are in place. It too is irreducibly complex.

Behe writes in a tremendously plausible sort of way, deftly using turns of phrase and attractive examples to bolster his case. I say, in admiration and without irony or sarcasm, that he must be a wonderful classroom teacher. Yet those with knowledge of the pertinent science have been all over the IDT claims, showing that far from inexplicable by selection, biologists now have some pretty good ideas about how these sorts of things occur.[14] A couple of things to keep in mind are that rarely if ever is a complex part started from scratch. As often as not, something being used for an entirely different process is co-opted and then put to use. As one example, take the Krebs cycle, a highly complex process with many steps, used by the cell to provide energy. It did not just spring into being. It was a "bricolage," built bit by bit from other pieces. The Krebs cycle was built through the process that Jacob called "evolution by molecular tinkering," stating that evolution does not produce novelties from scratch: it works on what already exists.[15] In any case, it is simply not true that the sorts of examples that Behe provides have no antecedents or clues as to how they might have come into being gradually. The blood-clotting cascade has over thirty moves, but many are more or less repetitions and could simply have come through duplication.[16] There are examples of functioning cascades with far fewer steps. And the mousetrap has given rise to many happy hours of tinkering. You can, it appears, make a functioning trap with only four parts, with only three parts, with only two parts, and even with only one part. Admittedly, it is not a great trap, but remember that natural selection does not demand excellence. Just doing better than competitors. Until someone can show how you can make a trap with no parts at all, a one-part trap looks like a pretty good option.

Guided Evolution

What of the attempt to put God directly into the historical process, seeing his hand as guiding the course of evolution, from the primitive up to the human? This was the claim of Gray and Wallace, and of the post-*Origin* Tennyson. In the "Higher Pantheism," a poem read in 1869 at the first meeting of the Metaphysical Society (a group of believers and skeptics who met to discuss issues of mutual interest), God is behind all of the actions of unbroken law.

> God is law, say the wise; O Soul, and let us rejoice,
> For if His thunder by law the thunder is yet His voice.
>
> Law is God, say some: no God at all, says the fool;
> For all we have power to see is a straight staff bent in a pool;
>
> And the ear of man cannot hear, and the eye of man cannot
> see;
> But if we could see and hear, this Vision—were it not He?

And, to the end of his life, Tennyson saw God as working his purpose out.

Where is one that, born of woman, altogether can escape
From the lower world within him, moods of tiger, or of ape?
Man as yet is being made, and ere the crowning Age of ages
Shall not aeon after aeon pass and touch him into shape?

All about him shadow still, but, while the races flower and
fade,
Prophet-eyes may catch a glory slowly gaining on the shade,
Till the peoples all are one, and all their voices blend in
choric
Hallelujah to the Maker "It is finish'd. Man is made."

The story continues down to the present. John Paul II, conservative about doctrine and morality, was ever sympathetic to science—a reflection perhaps of the fact that he was the most famous professor from Cracow University in Poland since Nicolaus Copernicus. He embraced evolution, even Darwinism. But blind law could not do it all. In his encyclical *Humani Generis*, Pope Paul VI asserted: "The spiritual soul is created by God." His successor reaffirmed this point: "As a result, the theories of evolution which, because of the philosophies which inspire them, regard the spirit either as emerging from the forces of living matter, or as a simple epiphenomenon of that matter, are incompatible with the truth about man. They are therefore unable to serve as the basis for the dignity of the human person"; adding, "with man, we find ourselves facing a different ontological order—an ontological leap, we could say."[17]

Protestants feel much this way too. The physicist-theologian Robert J. Russell calls his position NIODA, "Non-Interventionist Objective Divine Action."[18] Russell invokes quantum theory, suggesting that perhaps God flies below the radar, as it were. Think about mutations. In the Darwinian picture they occur, often on a regular basis, but have no direction. There are various reasons why they occur. Miscopying is a popular cause. We can even tell such things as why miscopying is more common in some cases than in others. At least one known cause of mutations occurs right down at the quantum level. Something happens and there is a knock-on effect and a new variation. But the thing about quantum events is that, although we may be able to quantify them over groups, we can never pin down any particular event. This is the crux. In time t, x% of genes A will mutate into genes B. Now it might be that ten seconds into t, a change would make no big difference, but that ten thousand seconds into t, a change might make all of the difference, because just such a mutation would then be needed and used. Russell suggests that this is where God makes his moves. From the viewpoint of modern science, he doesn't interfere—you still get the same x% in time t—but when the actual mutation occurs is crucial, and God is in charge here. So God could and does guide evolution to the production of human beings.

The philosopher Elliott Sober thinks along somewhat similar lines.[19] He is a nonbeliever so stresses that he does not think that God does intervene. Just that there is nothing in evolutionary theory that says he cannot intervene. In respects, it seems as if Sober's position is even stronger than Russell's because he does not need quantum mechanics to do the trick. No theory logically precludes some other (as yet unknown) natural cause influencing mutation rates—heat might be ignored but could be a factor—so God could step in and alter things, thus countering the unknown factor(s), and so we remain in ignorance. Or perhaps we note the changes but still have no right to insist on God. There could be something else, natural, in play. Staying with Russell, who believes that God does intervene, he thinks that his position (NIODA) is far superior to IDT. Perhaps so. It seems to me that will still have horrendous theological difficulties. If God was prepared to mutate the sperm of some poor little monkey so it could end up as our great grandfather, why does he not mutate the sperm of some chap who is about to parent a child with grave genetic issues? Once God starts to get involved in the processes, it never ends. But whether or not you agree that you should keep him out from the start, the fact is that John Paul II and Robert J. Russell are part of a tradition that thinks history shows purpose, the evolution up to humankind, and that the only way in which this could have occurred is by the direct designing intervention of the Almighty.

The Anthropic Principle

One thing that has rather dropped out of the conversation was something Plato thought an important part of the story, namely, purpose in the nonliving world. He (and Aristotle) thought we see design and purpose there, if not (at least for Aristotle) as strongly as in the living world. Although we shall see some post–Scientific Revolution support for final-cause thinking in the nonliving world, in optics particularly, generally such talk was absent and if promoted, strongly frowned upon. Remember the story of the chemist James Lovelock and the biologist Lynn Margulis—two very good scientists—who conceived the Gaia hypothesis, the idea that the earth is a living organism?[20] They argued that this is shown by the fact that the earth's temperature and atmosphere has remained far more stable—it is "homeostatic"—than one might have expected by chance, and that this demonstrates its organic nature. Things happen in or on the physical world to promote its well-being. Atmospheres and seas and the like function as they do in order to maintain stability. The outcry was deafening, with scientists scrambling to get in their criticisms. Typical was the reaction of the Canadian molecular biologist Ford Doolittle that although Jim Lovelock's "engaging little book" gives one "a warm comforting feeling about Nature and man's place in it," it is based on a view of natural selection that "is unquestionably false."[21] Others were significantly less courteous.

It was biologists who reacted most strongly against Gaia, because it is they who have been fighting the purpose wars for so long, and who feel now—thanks to natural selection—they are finally winning the battle. Generally, physicists stayed out of the fight, which perhaps prepares us for the fact that today there are some physicists and fellow travelers—philosophers and theologians—who want to revive the argument from design, now basing their claims on the physical world rather than the organic world. Discussion centers on the "Anthropic Principle," something said to come in a number of versions.[22] Most obvious is the "Weak Anthropic Principle," the WAP: "The observed values of all physical and cosmological quantities are not equally probable but they take on values restricted by the requirement that there exist sites where carbon-based life can evolve and by the requirements that the Universe be old enough for it to have already done so."[23] This does not say a huge amount and is fairly uncontentious. If you have life like we have, then the conditions for it must have been such that life like we have appears and is sustainable.

More interesting is the "Strong Anthropic Principle," SAP. The key idea here is that the universe had to be "fine-tuned" to get life going at all and sustain it. The various constants that govern the laws of nature could not be chosen at random but had to be very exact within incredibly narrow limits. "There exists one possible Universe 'designed' with the goal of generating and sustaining 'observers.' "[24] In other words, the lack of randomness implies a designer of some sort. What constants are we thinking of? Gravity for a beginning. It is $1,0^{39}$ times weaker than electromagnetism. Which is just as well, for if gravity had only been $1,0^{33}$ times weaker than electromagnetism, the suns of the universe would be a billion times less big and burn a million times faster. Analogously, the nuclear weak force is $1,0^{28}$ times weaker than gravity. If it had been slightly weaker, the hydrogen of the universe would have been converted to helium, and that would have meant no water. Life as we know it would not be possible.

All of this points at least to an updated version of the Demiurge, even if (as is the case with design arguments) it cannot take us all the way to a Creator God. Plato was right after all! Or was he? Part of the trouble with these anthropic arguments is that we are working from a single example—our world—and it is so difficult to know if we are unique or what might have happened pretty much all of the time. Think of a number. Double it, and the answer you want is a half. Suppose there really are multiverses, alternative universes, as many physicists believe? Wouldn't a universe like ours be bound to crop up if you tried enough times? Physics Nobel Prize winner Steven Weinberg writes:

> In any such picture, in which the universe contains many parts with different values for what we call the constants of nature, there would be no difficulty in understanding why these constants take values favorable to intelligent life. There would be a vast number of big bangs in which the constants of nature take values unfavorable for life, and many fewer where life is possible. You don't have to invoke a benevolent designer to explain why we are in one of the parts of the universe where life is possible: in all the other parts of the universe there is no one to raise the question.[25]

The fact that it is our universe or part of the universe in which there is life is no more improbable than that someone holds a winning lottery ticket. Given enough rolls of the dice or the drum, there was bound to be a winner eventually. The same with livable universes. Obviously, if we didn't hold the winning ticket, we wouldn't be around to tell everyone about it. That's no miracle, any more than that the person who won the lottery is the person who quits work and goes to live in the South of France.

Even in our universe, one gets the sense that often these anthropic arguments work from the alteration of just one parameter.[26] Everything collapses, and the cry is that there must be more than chance. But what if you alter not just one constant but several in unison? It is less obvious now that life is impossible. Think of an analogy. You have a soccer team with a brilliant center forward. Your whole strategy is built on getting him and the ball up close to the opponents' goal while avoiding the offside rule. Then he breaks a leg. Does this mean you will never win another game? No! However, you probably aren't simply going to substitute for him, using someone else in the same role. You might, for instance, start to pay more attention to defense, hoping that the opposition will wear down and then collectively you can strike. As in physics. Pushing the analogy a bit, if your ultimate aim is to make a living entertaining spectators, you might use your team's talents by switching from soccer to cricket or baseball, and providing thrills there instead. Are we convinced that only the kind of life-form we know—carbon-based and so forth—is the only viable life-form? What about the Horta in *Star Trek*? Unlike some people I know, I am not that keen on sex with someone made from silicone, but if they can write music like Bach, I'm game. To go to the concert, that is.

There are still the Humean philosophical arguments, holding as much here as with traditional design arguments. Do we have one designer or a squad of designers? Is there a trail of botched attempts and are we just one attempt on a course to a perfect universe? What about the problem of evil? Is this something that had to be? And so forth. And in any case, Weinberg suggests that often things are not as precise—nor need they be as precise—as people think. The formation of carbon is often cited as a case where super accuracy was needed. To make carbon from helium, you need a huge energy state above normal—in fact, about 7 million electron volts (MeV) above normal. However, it also turns out that if you go over 7.7 MeV, things won't work properly. Miraculously, apparently, there is such a needed state for carbon, which comes in at 7.65 MeV. All of this surely cannot be chance. Carbon misses the cutoff by .05 MeV, or less than 1 percent. Over the cutoff and no life. Under the cutoff and abundant life. But, asks Weinberg, why the figure of 7.65? It turns out that this is a function of the carbon production—first you combine two helium nuclei to make beryllium, and then you bring in a third to make carbon. And here, apparently, there is more flexibility. The beryllium-helium join up occurs at 7.4 MeV. So you cannot go more than 0.3 MeV more before things come apart. But this means that although, in fact, the carbon state (at 7.65) misses the target by .05 MeV, remember: this is against a rise of .25 MeV (from 7.4 to 7.65). This means that the upper target of 7.7 is actually missed by 20 percent (.05/.25) rather than 1 percent. So, in other words, while it is true that the 7.7 figure is fixed as we know it, further down the road, things are nothing like as tight, and the coincidences seem less striking.

There are other cases where the "coincidences" involve orders of magnitude of wriggle room.[27] No need to pursue things here, for in the end, the trouble with these sorts of arguments is that nobody is going to change anyone else's mind. Biologists are so sick of design arguments that if the heavens opened and Almighty God yelled down that he exists, they would not take him seriously. Physicists are to the sciences what philosophers are to the humanities, convinced that they are the best and brightest, and no one has any authority to challenge them. Even if they are not themselves very keen on design arguments, they are not about to let biologists make the decisions. Let us simply conclude that one should beware of the gods bearing gifts. Anthropic arguments have a way to go before they will be at all convincing, and the science is developing so quickly that in the end there may indeed be nothing to grasp.

Concluding Unscientific Postscript

One final word. God had a rough time in the nineteenth century. For all that, there were those who saw in the devastation ways of rebuilding in a more theologically satisfying manner. The arguments for the existence of God—particularly the argument from design—were taking a horrendous beating. Perhaps the right way forward was not to defend them—as in this chapter we have seen people doing and as, in the religious world, Thomists still do—but to argue that the demise of natural theology was a good thing. "Teleological observations on things often proceed from a well-meant wish to display the wisdom of God as it is especially revealed in nature . . . Whole books used to be written in this spirit. It is easy to see that they promoted the genuine interest neither of religion nor of science. External design stands immediately in front of the idea: but what thus stands on the threshold often for that reason is least adequate."[28] Reason undermines faith. True belief in God—faith-based belief in God—is meaningless, or at the least gelded, by an underpinning of argument. Wherein lies the virtue if all you are doing is that which can be demonstrated? In his analysis of the Abraham and Isaac story, the Danish philosopher-theologian Søren Kierkegaard stresses how Abraham's action of taking his son to the altar to be sacrificed is truly "absurd."[29] And yet it is the paradigmatic example of faith. In the twentieth century, this idea was picked up and strongly endorsed and defended by the Swiss theologian Karl Barth: "Every visible status, every temporal road, every pragmatic approach to faith, is, in the end, the negation of faith."[30]

Whether or not one can or should thus eliminate natural theology, including design arguments, has been happy fodder for many a student in search of a topic for a doctoral dissertation. Remember: "The heavens declare the glory of God; and the firmament sheweth his handiwork" (Ps. 19:1). The point is that in the eyes of many sophisticated Christians, wanting to make the second part of the Design argument, from design-like to real Design, is false theology. Battles over Intelligent Design Theory and the anthropic principle and the like have a very old-fashioned look about them. A bit like arguing about whether it is moral for women to use the pill.

CHAPTER EIGHT

Aristotle Redivivus

Aristotle Lives On

During and after the Scientific Revolution, the personification of nature that is at the heart of the Aristotelian philosophy had a nasty way of reappearing in the most orthodox of machine-metaphor-influenced places. Take Galileo's *Two Dialogues*, surely the poster child of the new mechanistic approach to nature. Yet consider: "it is as though we have been led by the hand to the investigation of naturally accelerated motion by consideration of the custom and procedure of nature herself in all her other works, in the performance of which she habitually employs the first, simplest, and easiest means. And indeed, no one of judgment believes that swimming or flying can be accomplished in a simpler or easier way than that which fish and birds employ by natural instinct."[1] If this isn't a teleological picture, with nature doing what is of value—doing things by the simplest means—it is hard to know what is.

Even more than mechanics, optics was riddled with final-cause thinking. Fermat's "principle of least time" explains Snell's "law of refraction," the connection between the angle of incidence and the angle of refraction. Since light going from a less dense to a denser medium is bent toward the normal, it is not going from beginning to end by the shortest distance. But assuming that light travels less quickly in a more dense than less dense medium, one can show that it does travel in the shortest time (because by being refracted, it minimizes the distance it has to go in the denser medium). What is this principle but an appeal to simplicity and value? In order to get from point A to point B as quickly as possible, it takes this route over all others. The purpose of taking this route is to be as quick as possible. Deliciously, even Descartes is into this kind of reasoning. "While Nature has several ways to arrive at an effect, she always infallibly follows the shortest."[2]

Is this genuinely neo-Aristotelian? No one back then was asking quite the questions we are asking now. Commenting on these issues, the French mathematician and philosopher Pierre Louis Maupertuis (1698–1759) worried: "I know the distaste that many mathematicians have for *final causes* applied to physics, a distaste that I share up to some point. I admit, it is risky to introduce such elements; their use is dangerous, as shown by the errors made by Fermat and Leibniz in following them." However, he consoled himself that in the end, God puts everything to rights!

> One cannot doubt that everything is governed by a supreme Being who has imposed forces on material objects, forces that show his power, just as he has fated those objects to execute actions that demonstrate his wisdom. The harmony between these two attributes is so perfect, that undoubtedly all the effects of Nature could be derived from each one taken separately. A blind and deterministic mechanics follows the plans of a perfectly clear and free Intellect. If our spirits were sufficiently vast, we would also see the causes of all physical effects, either by studying the properties of material bodies or by studying what would be most suitable for them to do.[3]

Is this Platonic, Aristotelian, or, perhaps most accurately, an Aristotelian picture that has behind it a Designer God rather than an indifferent Unmoved Mover? If this last is the option, then one can see how the Designer God could get ever-more remote with only the picture remaining. What is important is that the picture not only remained right through the heyday of mechanism but flourished! And as the eighteenth century went on, more and more serious thinkers felt the case was building. You should remember that this is just the time when people like Benjamin Franklin were pushing the science of electricity forward, suggesting that it is forces or invisible fluids at issue here, and others were showing that life itself (obviously) involves not just fluids but electrical discharges, as are needed for the functioning of muscles. Even physics was being co-opted into this movement, for the Jesuit Roger Boscovich argued—in a move endorsed by Kant in his *Metaphysical Foundations of Science*—that matter itself can be reduced to opposing forces, those pushing out and those pulling in. It pushes out as you try to penetrate it, but at the same time it pulls back in, or it would simply diffuse throughout the universe: "repelling force belongs to the essence of matter as much as attractive force does—the two can't be separated in the concept of matter."[4]

Self-Organization

This is not Cartesian *res extensa*, at least not what we thought was meant by Cartesian *res extensa*, a sentiment endorsed by Spinoza. He spoke in a monistic fashion of *Deus sive Natura*. Somehow, the whole of nature is divine and living: "Besides God no substance can be granted or conceived."[5] Perspectives like this found increasing support as the eighteenth century drew to an end and, in reaction to mechanism (and related phenomena like the Industrial Revolution), there was the growth of the philosophy and school—Romanticism—that made feeling, emotion, and life absolutely central.[6] One who was deeply influenced by Spinoza was the major philosophical successor to Kant, Friedrich Wilhelm Joseph [von] Schelling (1775–1854). He saw that attempts to divide knowledge into subjective and objective are as doomed to failure as attempts to separate the animate from the inanimate. He quoted approvingly from Spinoza's masterwork, the *Ethics*, "that whatsoever can be perceived by the infinite intellect as constituting the essence of substance, belongs altogether only to one substance: consequently, substance thinking and substance extended are one and the same substance, comprehended now through one attribute, now through the other."[7] At the same time, there was the influence of Plato—as a teenager, Schelling penned a sixty-page essay on the *Timaeus*—and so the world in any sense must be essentially organic, with final cause an essential part of it. "Even in mere organized matter there is *life*, but a life of a more restricted kind. This idea is so old, and has hitherto persisted so constantly in the most varied forms, right up to the present day—(already in the most ancient times it was believed that the whole world was pervaded by an animating principle, called the world-soul, and the later period of Leibniz gave every plant its soul)—that one may very well surmise from the beginning that there must be some reason latent in the human mind itself for this natural belief."[8] There is indeed a reason for this belief. "The sheer wonder which surrounds the problem of the origin of organic bodies, therefore, is due to the fact that in these things necessity and contingency are most intimately united. *Necessity*, because their very *existence* is *purposive*, not only their form (as in the work of art), *contingency*, because this purposiveness is nevertheless actual only for an intuiting and reflective being."[9]

"Their very existence is purposive"! Schelling highlights an earlier term that is very popular in some circles today. Apparently, because of the notion of purpose, "the human mind was very early led to the idea of a *self*[-]organizing matter, and because organization is conceivable only in relation to a mind, to an original union of mind and matter in these things. It saw itself compelled to seek the reason for these things, on the one hand in Nature itself, and on the other, in a principle exalted above

Nature; and hence it very soon fell into thinking of mind and Nature as one."[10] Self-organization! The world is something that produces itself, has its developing powers inside, as an unfurling organism is driven by forces within rather than without. "Nature should be Mind made visible, Mind the invisible Nature. Here then, in the absolute identity of Mind *in us* and Nature *outside us*, the problem of the possibility of a Nature external to us must be resolved. The final goal of our further research is, therefore, this idea of Nature; if we succeed in attaining this, we can also be certain to have dealt satisfactorily with that Problem."[11]

There is more we could say about this vision, which, translated into science, became known as Naturphilosophie or "Nature Philosophy"—for instance, about the work and influence of the poet Johann Wolfgang von Goethe. He devoted many of his formidable powers to the study of nature—taking on mechanism full front, especially in his attack on the Newtonian theory of light—and in the biological world, explicitly endorsing an organic model that owed much to Greek thought. But seizing on the Aristotelian notion of self-organization, jump a century and move to Scotland, and consider the work of the morphologist D'Arcy Wentworth Thompson, *On Growth and Form*, first published in 1917.[12] Significantly, championed by Gould, himself then working up to a sustained attack on Darwinism,[13] Thompson had little time for natural selection or for the whole tradition that it represented. He always looked back beyond the Enlightenment and the two thousand years leading up to it, finding his true spiritual home back in ancient Athens. Like Aristotle particularly, he was ever committed to a world that was more than just dead matter, a world that in some sense was living with the consequent absolute value that that implied.

> The waves of the sea, the little ripples on the shore, the sweeping curve of the sandy bay between its headlands, the outline of the hills, the shape of the clouds, all these are so many riddles of form, so many problems of morphology, and

> all of them the physicist can more or less easily read and adequately solve: solving them by reference to their antecedent phenomena, in the material system of mechanical forces to which they belong, and to which we interpret them as being due. They have also, doubtless, their immanent teleological significance; but it is on another plane of thought from the physicist's that we contemplate their intrinsic harmony and perfection, and "see that they are good."[14]

Thompson adds at once, "Nor is it otherwise with the material forms of living things."

Thompson was totally committed to function, to purpose—he is not denying the first part of the Design argument, about the design-like nature of organisms—but he was as totally committed to the idea that ends, values, emerge spontaneously from the physicochemical workings of the world, without need of natural selection, which later he saw as having only the negative end of removing inadequate or failing forms. This neo-Aristotelian thinking comes through clearly in his discussion of the forms of jellyfish. A Darwinian would at once look for function and why selection might have favored one form rather than another. Thompson equally looked for function but saw it all as a matter of the physics of denser fluids in less dense fluids. "To let a drop of ink fall into water is a simple and most beautiful experiment. The effect is more violent than in the former case. The descending drop turns into a complete vortex-ring; it expands and attenuates; it waves about, and the descending loops again turn into incipient vortices."[15] Continuing, that "instead of letting our drop rise or fall freely, we may use a hanging drop, which, while it sinks, remains suspended to the surface. Thus it cannot form a complete annulus, but only a partial vortex suspended by a thread or column—just as in Overbeck's jet experiments; and the figure so produced, in either case, is closely analogous to that of a medusa or jellyfish, with its bell or 'umbrella,' and its clapper or 'manubrium' as well."[16]

Function, purpose, final cause, values: "When, after attempting to comprehend the exquisite adaptation of the swallow or the albatross to the navigation of the air, we try to pass beyond the empirical study and contemplation of such perfection of mechanical fitness, and to ask how such fitness came to be, then indeed we may be excused if we stand wrapt in wonderment, and if our minds be occupied and even satisfied with the conception of a final cause."[17] Just not from natural selection, but as he basically admitted, somehow from the very workings of nature itself. "And yet all the while, with no loss of wonderment nor lack of reverence, do we find ourselves constrained to believe that somehow or other, in dynamical principles and natural law, there lie hidden the steps and stages of physical causation by which the material structure was so shapen to its ends."[18]

Due in no small part to the coming of computers, there is now a whole school that works in Thompson's tradition, trying to show how features Darwinians ascribe to selection are truly the result of mathematics and nature's unguided laws. Perhaps a function of the hostility to the perceived blind ruthlessness of Darwinian selection, there is often—as in the case of Gould—some ambiguity as to whether the claim is (as it was for Thompson) that ends are being served by physical law unaided, or if the very urge for ends is itself being denied or downgraded. Is the first part of the Design argument being accepted or rejected? The slogan of this school—"order for free"—suggests that there is something to do with purposes, something of value, but whether this something is a utilitarian end or just an elegant pattern is often left hanging.[19]

Phyllotaxis, something picked up by Thompson, has been a favorite topic. In many plants—sunflowers are the paradigmatic example—the seeds are packed in a very distinctive manner, showing different curves and spirals. One can readily show that the pattern is susceptible to mathematical analysis and somewhat remarkably is found to be determined by Fibonacci sequences. These are mathematical formulae made famous by the

Da Vinci Code, where each number in a sequence is the sum of the numbers preceding it. Thus, 0, 1, 1, 2, 3, 5, 8, 13, and so on. Evolutionists have long known about this, and Asa Gray seized on it as an example of selection in action,[20] distributing seeds and other plant parts to their advantage—for instance, in dispersal. Those in the Thompson tradition argue that it is simply a mathematical artifact of the way in which seeds and plant parts are produced, from the center out, and no more—or less—should be read into this. Obviously they think it but the tip of a very large iceberg. The late Brian Goodwin, a lifelong maverick, Canadian-born morphologist, was practically Pythagorean in his numerological enthusiasms. The vulgar fraction series formed by dividing successive members of the Fibonacci series homes in on 0.618, which in turn is what the ancient Greeks called the "golden mean," the figure arrived at by dividing the sides of a rectangle such that removing a square from the rectangle leaves one with a smaller but identical rectangle. As it happens, you can get the golden mean out of circles also, if you divide up the perimeter properly. This gives you a major angle of 137.5 degrees, which (and if you are not yet convinced you will be now!) is just the angle on the genetic spiral that divides successive leaves or parts. "So plants with spiral phyllotaxis tend to locate successive leaves at an angle that divides the circle of the meristem in the proportions of the Golden Section. Plants seem to know a lot about harmonious properties and architectural principles."[21] (The meristem is the growing tip of the plant.)

And at this point, perhaps expectedly, values make their appearance. Goodwin has nothing but contempt for a philosophy that attempts to take meaning and value out of existence. Everything is interconnected, in an essentially harmonious fashion, with shared values. Darwinism is "an extreme reductionism that makes it impossible for us to understand concepts such as health. Health refers to wholes, the dynamics of whole organisms. We currently experience crises of health, of the environment, of the community. I think they are all related. They are not caused by biology by any means, but biology contributes to these crises by failing to give us adequate conceptual understanding of life and wholes, of ecosystems, of the biosphere, and it's all because of genetic reductionism."[22] We have got to escape the Darwinian metaphors of "competition and conflict and survival," replacing them with metaphors stressing organisms as "co-operative as they are competitive." We must turn from "nature red in tooth and claw, with fierce competition and the survivors coming away with the spoils." We need a new perspective where the "whole metaphor of evolution, instead of being one of competition, conflict and survival, becomes one of creativity and transformation."

This is not a man—or school—that has turned his back on purpose. It is all a question of which purposes and how to get them.

Vital Forces

With Goodwin, we are moving from the individual organism and its purposes to the group and beyond to history and its purposes. Along with the Aristotelian approach to the individual, there was also an Aristotelian approach to history, or perhaps more accurately, since Aristotle himself was not into this kind of inquiry, an Aristotle-inspired approach to history. We find this in Lamarck.[23] The mechanism given his name, the inheritance of acquired characteristics, was always secondary for him. Primary was a kind of upward force, from the spontaneously generated to our species, not in a treelike-branching fashion made famous by Darwin but in parallel lines of ascent, going through the same stages, some having started earlier than others.[24] Hence (say), were lions to go extinct, more would be on their way, later. He wrote of a "life force," "le pouvoir de la vie." This seems to lead to complexity, and then "Lamarckism" tones things up adaptively.

In Germany we find similar thinking, although often idealistic rather than materialistic. Thus Hegel: "Nature is to be regarded as a system of stages, one arising necessarily from the other and being a proximate truth of the stage from which it results; but it is not generated naturally out of the other, but only in the inner Idea which constitutes the ground of Nature."[25] This sort of thinking crossed the Channel and, through the medium of the poet Samuel Coleridge, who virtually plagiarized the writings of Schelling, had a huge influence on Herbert Spencer, the evolutionist who, in respects, had even more effect on the general public than did Darwin.[26] Spencer saw organic evolution as being but one facet of the overall upward progress that characterizes the whole world process: from the undifferentiated to the differentiated, or in his words, from the homogeneous to the heterogeneous:

> Now we propose in the first place to show, that this law of organic progress is the law of all progress. Whether it be in the development of the Earth, in the development of Life upon its surface, in the development of Society, of Government, of Manufactures, of Commerce, of Language, Literature, Science, Art, this same evolution of the simple into the complex, through successive differentiations, holds throughout. From the earliest traceable cosmical changes down to the latest results of civilization, we shall find that the transformation of the homogeneous into the heterogeneous, is that in which Progress essentially consists.[27]

Nothing escapes this law. Humans are more complex or heterogeneous than other animals; Europeans are more complex or heterogeneous than savages; and (hardly a surprise) the English language is more complex or heterogeneous than the languages of other speakers.

Eclectically grabbing bits and pieces from everywhere, Spencer propounded his theory of "dynamic equilibrium."[28] Societies are like organisms.[29] Every now and then they get disturbed and then they strive to reachieve equilibrium, but at a higher, more differentiated level. It is not quite obvious why this happens, but that causes have a natural tendency to produce multiple effects and hence complexity is important. Whatever the case may be, it is a better state to which we are ever pointed, and this is why morality is essentially a function of aiding the processes of evolution. "Ethics has for its subject-matter, that form which universal conduct assumes during the last stages of its evolution."[30] Continuing: "And there has followed the corollary that conduct gains ethical sanction in proportion as the activities, becoming less and less militant and more and more industrial, are such as do not necessitate mutual injury or hindrance, but consist with, and are furthered by, co-operation and mutual aid."[31] Although like many Victorians of his age, Spencer was very critical of conventional religion and inclined to think of himself as an agnostic, he would talk of the Unknowable almost as if it were an entity rather than a confession of ignorance: "the Power which the Universe manifests to us is utterly inscrutable." All of this—especially the built-in upward progress—inclines one to think that for Spencer there is more than brute fact and perhaps at the least some kind of objective principle of ordering, a kind of principle of purpose to the order of history.

Moving back to the Continent, there were the so-called vitalists. From Germany, the embryologist Hans Driesch (1908), who supposed the notion of an *entelechy*, is best known.[32] From France, we have already met the philosopher Henri Bergson, who supposed the notion of an élan vital. Much influenced by Herbert Spencer, in *Creative Evolution*—the very title is suggestive—Bergson made it clear that all of life is bound together, and the course of history is not random but pointed to the emergence of humankind. "Where, then, does the vital principle of the individual begin or end? Gradually we shall be carried further and further back, up to the individual's remotest ancestors: we shall find him solidary with each of them, solidary with that little mass of protoplasmic jelly which is probably at the root of the genealogical tree of life." Continuing:

> In this sense each individual may be said to remain united with the totality of living beings by invisible bonds. So it is of no use to try to restrict finality to the individuality of the living being. If there is finality in the world of life, it includes the whole of life in a single indivisible embrace. This life common to all the living undoubtedly presents many gaps and incoherences, and again it is not so mathematically *one* that it cannot allow each being to become individualized to a certain degree. But it forms a single whole, none the less; and we have to choose between the out-and-out negation of finality and the hypothesis which co-ordinates not only the parts of an organism with the organism itself, but also each living being with the collective whole of all others.[33]

Although I suggested earlier that the élan vital has an ontological presence—it is a physical thing in some sense, alien to Aristotle—overall with Bergson you don't get much more Aristotelian than that. Expectedly, like most philosophers, Bergson is pretty keen on having his cake and eating it too. On the one hand, he repudiates mechanism. Darwinism gets short shrift. On the other hand, he does not want the whole of life predetermined by an outside plan, allowing no room within for choice or creativity. That said, humans had better come out on top! Fortunately, they do: "not only does consciousness appear as the motive principle of evolution, but also, among conscious beings themselves, man comes to occupy a privileged place. Between him and the animals the difference is no longer one of degree, but of kind."[34] We can put matters more strongly: "in the last analysis, man might be considered the reason for the existence of the entire organization of life on our planet."[35]

In the second decade of the twentieth century, Bergson was immensely popular. I note with some ironic amusement, given how Richard Dawkins has so enthusiastically endorsed the idea, that Julian Huxley was an ardent Bergsonian and his introduction of arms races was intended to give a mechanistic, Darwinian backing to the French philosopher's overall vision. His friendship with the Huxleys (Julian and Aldous) was probably the conduit that informed D. H. Lawrence, and the reason why his two great novels, *The Rainbow* and *Women in Love*, are Bergsonian through and through. As I said, there are masses of rather silly talk about the power of blood—"A little flicker of rage ran through his blood. It was as if she were rousing him, goading him."—and so on and so forth. Silly, until one realizes that this is a metaphor for the élan vital. Listen, at the end of the second book, to the reflections of the hero Rupert Birkin.

> If humanity ran into a CUL DE SAC and expended itself, the timeless creative mystery would bring forth some other being, finer, more wonderful, some new, more lovely race, to carry on the embodiment of creation. The game was never up. The mystery of creation was fathomless, infallible, inexhaustible, forever. Races came and went, species passed away, but ever new species arose, more lovely, or equally lovely, always surpassing wonder. The fountain-head was incorruptible and unsearchable. It had no limits.[36]

Since Princeton University Press is a family-friendly organization, I will not go into the details of how Lawrence thinks this is to be achieved. I will simply say that Rupert's girlfriend, Ursula, has what might be described as a "multipurpose" body.

Julian Huxley was a lifelong enthusiast for Bergsonian vitalism. Even in his magisterial overview of neo-Darwinism, *Evolution: The Modern Synthesis*,[37] he admitted a fondness for the Bergsonian vision. In France, Pierre Teilhard de Chardin, probably in the first part of the twentieth century the country's best paleontologist, was likewise an enthusiast, linking Bergson's creative evolution with a Lamarckian upward progress to the so-called noosphere (the realm of human culture) with the Omega Point at the end, something that Teilhard (a Jesuit) identified with Jesus Christ.[38] Although an atheist, Huxley was the president of the British Teilhard de Chardin society. (Dobzhansky was the president of the American society.) Not that this brought Huxley much relief or relaxation. Reviewing Teilhard's masterwork, *The Phenomenon of Man* (1959), future Nobel Prize winner Peter Medawar wrote of "a feeling of suffocation, a gasping and flailing around for sense"; "a feeble argument, abominably expressed"; "the illusion of content"; and "alarming apocalyptic seizures."[39] And that was just the first paragraph. The real object of his scorn was Huxley, who had written a friendly introduction. Directly against Teilhard was Gould who, disliking intensely the thoroughly progressivist nature of the *Phenomenon of Man*, accused the priest of having coordinated the Piltdown Hoax—a charge that was received with the contempt that it merited.[40]

Alfred North Whitehead

One of the more interesting figures on the intellectual scene of the early twentieth century was Alfred North Whitehead (1861–1947). He started life in England as a mathematician, with (his student) Bertrand Russell writing *Principia Mathematica*, a heroic attempt to show that mathematics is the deductive consequence of logic.[41] Late in life, he crossed the Atlantic and became professor of philosophy at Harvard, the joke being that when he entered the classroom, it was probably the first time in his life that he had ever attended a philosophy lecture. He set about formulating a metaphysical system, the very opposite from what one might expect from one with a grounding in mathematics. Whitehead's thinking, known aptly as "process philosophy," is deeply organic and developmental. In lectures given in 1925, he bemoaned the nature of materialistic evolution, arguing that it "is reduced to the role of being a word for another description of the changes of the external relations between portions of matter." Hence: "There can merely be change, purposeless and unprogressive." We seek a more creative, more dynamic process, explaining "the evolution of complex organisms from antecedent states of less complex organisms." Given this: "The doctrine thus cries aloud for a conception of organism as fundamental for nature. It also requires an underlying activity—a substantial activity—expressing itself in individual embodiments, and evolving in achievements of organism. The organism is a unit of emergent value, a real fusion of the characters of external objects, emerging for its own sake." Concluding, and making explicit our interests, "in the process of analyzing the character of nature in itself, we find that the emergence of organisms depends on a selective activity which is akin to purpose."[42]

It hardly needs remarking that none of this has anything to do with Darwin's theory of evolution by natural selection. There is clearly some affinity with some of the Bergson-like philosophies of creative evolution of the day—positions like that of Samuel Alexander, which saw levels of existence and new entities or wholes "emerging" at ever-higher levels.[43] But, truly, in seeking roots, one does better to go back to influences Whitehead notes, particularly Romanticism—"a protest on behalf of an organic view of nature, and also a protest against the exclusion of value from the essence of matter of fact."[44] In support, poets like Shelley ("Mont Blanc") are quoted:

> The everlasting universe of things
> Flows through the mind, and rolls its rapid waves,
> Now dark—now glittering—now reflecting gloom—
> Now lending splendour, where from secret springs
> The source of human thought its tribute brings
> Of waters—with a sound but half its own,
> Such as a feeble brook will oft assume,
> In the wild woods, among the mountains lone,
> Where waterfalls around it leap for ever,
> Where woods and winds contend, and a vast river
> Over its rocks ceaselessly bursts and raves.

Poets like Wordsworth and Shelley "express more clearly a feeling for nature, as exhibiting entwined prehensive unities, each suffused with modal presences of others."[45] Interpreting this gnomic judgment is left as an exercise for the reader.

Well known is the fact that Whitehead had far greater influence on American theology than American philosophy. In the hands of followers like Charles Hartshorne, process theology stressed the notion of "kenosis," that God (through Jesus) relinquished powers of divinity, and now is an evolving God involved in the world, laboring alongside his creation humankind. The poet Pattiann Rogers expresses some of these thoughts in "The Possible Suffering of God during Creation." Especially, there is the fear that it might not be worth the effort.

> Maybe he wakes periodically at night,
> Wiping away the tears he doesn't know
> He has cried in his sleep, not having had time yet to tell
> Himself precisely how it is he must mourn, not having had
> time yet
> To elicit from his creation its invention
> Of his own solace.[46]

Obviously the God of process theology is far from the God of Saint Augustine. His very being commits just about every heresy in the book. For all this, Hartshorne thought of him as a Platonic God, somewhat akin to the world soul—"the world is God's body." This sounds much like pantheism, Spinoza's *Deus sive Natura*, and—process people are into this—he invented his own fancy term, "panentheism." Whitehead rather nastily dissociated himself from this view: "In the *Timaeus* the doctrine [of the world soul] can be read as an allegory. In that case it was Plato's most unfortunate essay in mythology. The World-Soul, as an emanation, has been the parent of puerile metaphysics."[47] God is not part of the world. He cannot be creator, in the sense of designer, because in a sense that is still going on. Given the organic metaphor, we do perhaps have more of an Aristotelian flavor of an ongoing process, a force or power of creation. He spoke of God's "primordial nature," being "the unlimited conceptual realization of the absolute wealth of potentiality," and of "the lure for *feeling*, the eternal urge of desire."[48] Perhaps we should not read too much into this. Whitehead does not give the impression of a man with the history of philosophy in his bones, as it were. His God is a long way from the Aristotelian Unmoved Mover.

American Evolutionism

Whitehead is a one-off. The English cuckoo in the American nest. Heathcliff in the Earnshaw family. Except, really, he is less of a disturbance and more a figure who went his own way, with his own following, not overwhelmingly in the philosophical world. Apparently, he has followers in the business administration orbit.[49] So really, it is no great surprise that the kind of developmental philosophy promoted by people like him, and Bergson a little earlier, resonated more broadly in American culture. Discreet about but very appreciative of Bergson was one of the towering figures of twentieth-century evolutionary biology, the American population geneticist Sewall Wright.[50] Mention has been made already of his picture, the "shifting balance theory of evolution," one that sees creativity in small populations brought on by genetic drift, and subsequent amalgamation of these populations and selection working to finalize things in the greater whole. Wright—who thought himself a panpsychic monist, using this in the sense of one who sees everything as in some sense living and conscious—appreciated the progressivism of Bergson and read this also into his own world picture. He saw organisms as sitting on an "adaptive landscape" with peaks and valleys, and with an ever-higher movement from one peak to another. "The present discussion has dealt with the problem of evolution as one depending wholly on mechanism and chance. In recent years, there has been some tendency to revert to more or less mystical conceptions revolving around such phrases as 'emergent evolution' and 'creative evolution.' The writer must confess to a certain sympathy with such viewpoints philosophically."[51]

More influential on Wright—although only confirming the neo-Aristotelian trend in his work—was Herbert Spencer. We should perhaps be primed for this, for if you look at the discussion of progress by Julian Huxley, it is at least as much Spencerian as Bergsonian. Remember, to requote: "it comes to pass that the continuous change which is passing that through the organic world appears as a succession of phases of equilibrium, each one on a higher average plane of independence than the one before, and each inevitably calling up and giving place to one still higher."[52] You don't get much more Spencerian than that, although perhaps the Bergsonian influence shows through (not in a contradictory manner) in Huxley's belief that the essential mark of progress is being independent of one's surroundings.

In the case of Wright, thanks to his father (who was one of his teachers), Spencer was an early influence and stayed with him. The very title of Wright's theory bringing in "balance" shows the trend of his thinking, and it was there throughout. Think of what happens. A population gets a shock or disruptive force breaking it up. Genetic drift occurs in the parts, making for more variation. The parts rejoin, with this new variation now part of the whole. And in the process, creative additions are made and the whole moves to a higher state. Dynamic evolution in Mendelian terms. "Evolution as a process of cumulative change depends on a proper balance of the conditions, which, at each level of organization—gene, chromosome, cell, individual, local race—make for genetic homogeneity or genetic heterogeneity of the species . . . The type and rate of evolution in such a system depend on a balance among the evolutionary pressures considered here."[53]

Wright had a huge influence on American evolutionary biology. How much the Spencerian influence continued it is hard to say. One, however, who felt the influence either directly from

Wright or perhaps in parallel from the men who taught Wright at Harvard in the 1910s—and who were the teachers of his teachers—was Edward O. Wilson. He has always been open in his admiration of Herbert Spencer, thinking him a much-unappreciated thinker. We shall see more of this when we come to the discussion of morality and its foundations. Here it is enough to note that Wilson's total conviction of the progressive nature of evolution and his total confidence that such a conviction needs little or no backing would fit nicely with someone whose metaphysics started with the belief that nature is not just dead molecules in motion but in some sense alive and directed toward ends—namely, human beings. Wilson has never read Aristotle, but the thread is there.[54]

We find traces (or more) of it also in other writings, particularly those who think that (in a kind of self-organizing way) nature is going to lead to complexity and hence to humankind. There are hints of this in some of Gould's later writings—he was often torn between denying progress and denying progress fueled by Darwinian factors. He certainly thought that nature had a natural tendency to complexity.[55] This occurs by a kind of random walk. You cannot get more simple than simple, but you can get more complex. A viewpoint endorsed—several times they acknowledge explicitly that they are standing in the tradition of Herbert Spencer—by Duke University colleagues, paleontologist Daniel McShea and philosopher Robert Brandon.[56] They promote what they proudly call "biology's first law." Named the "zero-force evolutionary law" or ZFEL, its general formulation runs: "In any evolutionary system in which there is variation and heredity, there is a tendency for diversity and complexity to increase, one that is always present but that may be opposed or augmented by natural selection, other forces, or constraints acting on diversity or complexity."[57] It is something apparently with the status in evolutionary biology of Newton's first law of motion—a kind of background condition of stability, even though somewhat paradoxically their law suggests perpetual motion.

Although the authors are fairly (let us say) generic on their understandings of complexity and diversity—number of parts, number of kinds—the claims made are grandiose if familiar. A little like Sewall Wright's balance theory, the natural tendency to complexity—parts tend to be added on—generates new organic variations and hence types, and so one gets a version of order for free. As is usual in these discussions, it is not always obvious whether the claim is that adaptation is created in this way or if adaptation is now irrelevant. Probably more the former: "We raise the possibility that complex adaptive structures arise spontaneously in organisms with excess part types. One could call this self-organization. But it is more accurately described as the consequence of the explosion of combinatorial possibilities that naturally accompanies the interaction of a large diversity of arbitrary part types."[58]

As seems customary when philosophers write about intelligent design theory, there is—at least from a Darwinian perspective—a depressing sympathy for the position. There is no outright subscription to a Designer, however: "The creationist intuition, shared by many onlookers to the debate, is that it is difficult to see how the intermediates to these complex structures could have been functional, and therefore how they could have arisen and been maintained by natural selection."[59] And in an equally depressing but familiar manner, they go on to say: "But we point out, there may be another route available as well. If novel part types are delivered in excess, as the ZFEL suggests, then the combinatoric possibilities could be vast, with the result that colloquial complexity could—like pure complexity—be easy. And the role of natural selection could be mainly negative, revealing colloquial complexity by subtraction."[60] Selection is thus downgraded to an eliminative status rather than creative. We are right back to a world where nature itself has its own ends, its own purposes. There is no need of outside help or interference. "The scope we claim for the ZFEL is immodestly large. The claim is that the ZFEL tendency is and has been present in the background, pushing diversity and complexity upwards, in all populations, in all taxa, in all organisms, on all timescales, over the entire history of life, here on Earth and everywhere."[61]

Phylogeny Recapitulates Ontogeny

Given that German Romanticism was so indebted to Plato, not surprisingly there was always much interest in showing that the world, the organic world particularly, was one or One, linked throughout by revealing patterns. There was appreciation of function, but—in the footsteps of Aristotle in translating this into the world of organisms—form was prior. Goethe writing on plants made much of the similarities between parts, and the morphologists, like Lorenz Oken, carried this idea into anatomy and also saw similarities—homologies—between different organisms. It was natural that this sort of thinking would be pushed into understanding development, and before Darwin we find people like Louis Agassiz, Swiss ichthyologist (and later Harvard professor), pushing a threefold parallelism: the development of the individual, the development through time of the whole of life, and the relationships between organisms today. The last—the "chain of being"—is a very old idea with roots as far back as Plato's *Republic*, although it was Aristotle (in his *History of Animals*) who developed the idea, putting animals above plants (on grounds of their being able to move and sense) and then graded animals on reproductive modes and possession of blood.[62] The Christianized version saw the chain start with the most primitive organisms, work up through humans, on to angels, and then God at the top. Development through time was not necessarily taken in an evolutionary sense—when it was, it was called "phylogeny"—but more of an unfurling of creation through time as revealed, increasingly, by the fossil record. Individual development, "ontogeny," was of much interest to biologists. Karl Ernst von Baer (1792–1876) was the landmark figure in making much sense of the findings. In the 1860s, Darwin's follower in Germany, Ernst Haeckel (1834–1919), in fact as much influenced by Romanticism as by anything to be found in the *Origin*, gave the whole picture an evolutionary interpretation, expounding his well-known (albeit much criticized for its exceptions) "biogenetic law"—"ontogeny recapitulates phylogeny"—meaning that by studying the development of the individual one could discern the history of the group.

Since all of these thinkers, Haeckel especially, thought in terms of biological progress, there was ever a tendency to see in the kind of momentum that one sees in individual development—from the egg to the hen, from the embryo to the full-grown dog—a kind of analogous momentum in life's history, as it unfurls progressively to its end. In a parody one might say that phylogeny recapitulates ontogeny. A modern-day representative of this kind of thinking is the well-known science writer Robert Wright. He wonders if we might see a kind of progress to humankind, akin to the development we see in the individual organism. Drawing attention to the fact that Darwinians (among whom he numbers himself) like Richard Dawkins and the philosopher Daniel Dennett agree that the development of the organism is as much a design-like aspect of life as are the adaptations of the adult, Wright hypothesizes:

> We understand the physical process by which an egg unfolds into a squirrel. Yet Dennett and Dawkins agree that, in the case of a squirrel, we still need an additional "special kind of explanation"—namely, an explanation for how there came to be squirrel's eggs that do this sort of remarkable unfolding. Well, I'm making a comparable claim about the first seeds of life on Earth—the original self-replicating material that unfolded into the whole biosphere. I'm saying this unfolding, and the product of this unfolding, have properties that should lead us to suspect there is a "special kind of explanation" for how these seeds came to be here in the first place; I'm suggesting that these seeds, like squirrel's eggs, may be a

> product of "design" and have some "purpose." In other words, I'm suggesting that the word "seed" may be apt in a pretty strict sense.[63]

Speaking on behalf of Dawkins and Dennett, and for other Darwinians like myself, this simply won't fly. First, the development of the individual was produced by natural selection, and although individual parts of history may be controlled by selection, as through arms races, there is no suggestion that the process as a sequential and integrated whole is controlled by selection. The history of life on Earth is not the result of many histories in the past that gave rise to the next one and finally to the one that produced us. In any case, the adult organism like the squirrel is an integrated, functioning whole. The world taken as an entity is not. Although I have much sympathy for the Gaia hypothesis—it really does take seriously issues like the threat of global warming—overall, it just doesn't work. It really isn't the case, for example, that lagoons trap seawater, evaporate it, and then suck the salt underground to keep the salinity of the sea in a stable state.[64] Mount Everest just doesn't have the relationship with the Canadian prairies that the heart has with the eyes—in the former case, they just are contingently on Planet Earth, whereas in the latter case they are working conjointly to the benefit of the whole organism. And if you take just the biosphere, even though one might get sociality at restricted levels, the struggle for existence sinks all thoughts of overall integration. No progress here I am afraid.

An End with No Purpose

Plus ça change, plus c'est la même chose. We have the same philosophical approaches to the same philosophical problems. All that changes is the science in which everything is dressed up. And perhaps the nature of the Unmoved Mover. Although Aristotle did away with the Demiurge, an efficient-cause God, he kept with a final-cause God, seeing that values make little sense without a valuer or valuers. We and all of creation strive to emulate It, because It is good. Neo-Aristotelians (unlike Neoplatonists) tend to make less of the gods or a God, but one senses that in some way they feel that human life in itself is of value and that nature strove to produce it because it was good. Herbert Spencer, never much given to modesty, false or otherwise, would probably have been happy to take on the divine role himself. Others, like Wilson, go more generally to humans as a whole, but ultimately it is we who make it all worthwhile. One doubts that Aristotle, with his keen sense of human fallibility, would have thought this quite enough.

CHAPTER NINE

Human Evolution

TAKE UP AGAIN our trichotomy—Plato, Aristotle, Kant. In respects, since these are such long-standing traditions, one expects to find merits in all three. Why else would they have persisted? In respects, since the science has moved on so dramatically—I refer now to Darwinian evolution through selection—and since the traditions were established before this great move, one expects to find none taken alone entirely adequate. This is true on both counts. Start with Plato. I myself have trouble with the Christian God and indeed with all and any gods. But this is by the by. The point here is that the Platonic tradition certainly captures important aspects of the forward-looking nature of the world. The catch is that in today's science—in today's Darwinian science—any kind of nonmechanical understanding is ruled out. You might think that God stands behind things, but that has to remain your opinion. So long as your deity does not flagrantly conflict with science, or so long as you think you can reconcile your deity with science—I for one am with that former professor at Calvin College in simply not seeing how you can simultaneously believe in an original Adam and Eve, crucial for the Augustinian position on original sin, with the history to be presented in this chapter—you are entitled to that opinion. Unfortunately, it isn't science, and you need to fill the gap now that God is no longer part of that picture.

Carry on with Aristotle. I have much sympathy for this position. If one means simply principles of ordering, then I am inclined to think that these properly supplement a Kantian position, explaining why it is that we can profitably think heuristically about ends and purposes. If it means anything more—as it clearly does to most of the people discussed in chapter 8, and as I am sure it does for Aristotle—then worries arise. If one makes Aristotle entirely secular—which he himself was not—then one runs afoul of this organic nature of reality. You have still got the job of explaining the principles. Is order for free really that plausible? Doesn't all experience point to the truth of Murphy's Law? If it can go wrong, it will go wrong. Without design or something equivalent, then nothing functions properly, and blind law without direction cannot do the job. Candidly, anyone who believes in ZFEL is probably ripe for Father Christmas. They think that lowering the taxes of the rich helps the poor. The world doesn't work this way, folks—it really doesn't. As Henry King discovered to his chagrin, string ties itself in knots. On its own, it doesn't coil up nicely.

Which brings us to the third option, the Kantian approach. This fits with a mechanistic view of reality, while at the same time taking purpose, teleology, as fundamental and irreducible. You are not going to get rid of it, nor should you. The problem is, why does this work? We know the answer. Natural selection! This produces design-like effects and yet is entirely mechanistic. It shows why Murphy's Law is not all-conquering. There is order in the world, produced by natural selection, and we react to it. Whether or not this means that there is or was progress up to human beings is at least debatable. Obviously, natural selection could produce humans because it has produced humans. The question is whether there is any necessity to this process or whether it was just accidental, be it one-off or repeated many times in this or other universes. Stress again, though, that there is nothing in our science or our philosophy that gives Meaning to the world. You might bring Meaning to the world, but that is another matter.

I come therefore to the rest of this book. Moving on now beyond the history of why we have arrived at this point, I want to explore what the Kantian/Darwinian perspective implies for, makes clear about, us humans. What does it mean for us as thinking beings? What does it mean for us as religious beings? What does it mean for us as philosophical beings, interested in knowledge (epistemology) and morality (ethics)? What does it mean for us as human beings, cast naked into the Darwinian world? It is to these questions that I turn, as background starting with what we know of our own history.[1]

Human Evolution

Mentioned already was that Charles Darwin was absolutely convinced of the fact of human evolution and as soon as he had discovered natural selection was applying it to our species, to our minds and powers of thought no less. However, in the *Origin* he was cautious, wanting first to get the main details of his theory laid out for all to see and only at the end pointing to the implications for humankind. This did not stop others from getting on the bandwagon, and although in the *Descent* Darwin had much to say that was both new and interesting—notably about sexual selection—by then he was entering an already well-plowed field. Naturally, the early parts of the book were concerned with making the straightforward case for human evolution, showing how it is reasonable to think—especially on the evidence of homologies—that we and the higher apes are close relatives and that we came jointly from organisms more primitive. "It is notorious that man is constructed on the same general type or model with other mammals. All the bones in his skeleton can be compared with corresponding bones in a monkey, bat, or seal. So it is with his muscles, nerves, blood-vessels and internal viscera. The brain, the most important of all the organs, follows the same law, as shewn by Huxley and other anatomists."[2] No one, then or now, thought that our ancestors are alive still. The point is that we did come from monkeys, which in turn came from other mammals.

The two most distinctive things about humans are our large brains and our bipedalism. There are other distinctive features, for instance, being continually sexually receptive, but other than for teenagers, these don't count quite as high as being able to think and to walk upright. Darwin seized on them and suggested (what we shall see we now believe on good evidence) that bipedalism came before the explosion of brain size. He also believed (what we shall see we now believe on good evidence) that human origins lay in Africa. This was a lot more controversial. Many people would have much preferred Asia. The Chinese may not be Europeans, but they are certainly a step up on Africans. What Darwin did not have was any good fossil evidence of human evolution. When people referred to the "missing link," it was always that empty gap people had in mind—a gap perhaps made even more obvious and pressing by the discovery in the early 1860s of what was to become to this day one of the most famous of all linking fossils, *Archaeopteryx*, the reptile with feathers. From discoveries made in Europe in the years before Darwin, people knew about Neanderthals, but although they fit nicely into conceptions—from their fossils they do seem rather brute-like—no one was really convinced that they were an entirely different species from *Homo sapiens*. Indeed, some were inclined to think that if you went to the west coast of Ireland, you might well find representatives.

It was not until the end of the century that Eugene Dubois, a Dutch doctor and paleoanthropologist—as students of human evolution are known—digging in the Far East, found an indubitable specimen of a protohuman.[3] *Pithecampothrus erectus* or "Java Man" really was more primitive, although today we show that it was not that different from us by including it in the same genus as ourselves, *Homo erectus*. One thing that did convince was its place of discovery. In South Africa in the 1920s, when Raymond Dart, another doctor, started pushing the claims of a fossil found down at the bottom of the continent, Taung Baby, he had considerable difficulty in persuading people that he had something significant.[4] His task was not made easier by the biggest fraud in the history of science, Piltdown Man, a supposed ape-human found in England in the second decade of the century, and only unmasked in the 1950s. Eventually Dart's discovery was appreciated for what it was—a protohuman (these are known as "hominins") sufficiently aged that it was given a new genus, *Australopithecus*.

Then in the 1950s and beyond, thanks particularly to the labors of the indefatigable Leakey family, the fossils started to pour forth from central Africa, establishing beyond a doubt the place of our origin. American researcher Donald Johanson was at the heart of what is probably the most exciting discovery of them all—he certainly thinks it was the most exciting discovery of them all—the little biped nicknamed "Lucy."[5] She lived in Ethiopia just over three million years ago and had a brain about the size of a chimpanzee's. Be careful to understand what this means. Her brain was about 450cc as opposed to ours, about 1200cc. (Neanderthals' are slightly larger!) It does not mean she had a chimpanzee brain. By the indentations on the insides of skulls, you can tell quite a lot about the brains themselves, and it seems clear she had a brain on the way to being human. She really was the paradigmatic missing link, especially when researchers discovered that although she walked upright, she was better adapted at climbing trees than we. Although Lucy—now classified as *Australopithecus afarensis*—is very important, it is improbable that she is literally our ancestor. We now know that evolution is much more given to divergence—producing bush-like phylogenies (histories)—than simple unilineal change from one form to another. So Lucy would have been one of a number of species or subspecies, very closely related, one of which led to us.

Although still somewhat controversial—causing at least as much public interest as Lucy—was the discovery in the East, on one of the islands of Indonesia, of *Homo floresiensis*, a small humanlike being that stood about three feet six, with a small brain but with relatively large teeth and feet, that was, naturally, at once christened the "hobbit."[6] What makes this little being so interesting is that apparently it was thriving in the last hundred thousand years, and, although current opinion is inclined to set the date back a bit, may well have gone extinct less than twenty thousand years ago. This means that it would have coexisted with modern *Homo sapiens*, although there is no reason to think that there was actual physical overlap.

Starting about fifty years ago the study of the past was transformed by the coming of molecular biology. Already people were using physicochemical techniques to provide absolute dates of events in the past—rates of radioactive decay could yield much that is simply not there in the fossil record. Then it was realized that rates of genetic change at the molecular level could reveal much about relationships that could not be discovered from the fossil record alone. This was due primarily to the neutral theory of evolution, the brainchild of the Japanese evolutionist Moto Kimura, who reasoned that whether or not genetic drift occurs at the physical (phenotypic) level, at the molecular level a great deal of change would be, as it were, beneath the radar of natural selection, and so this essentially random change could be used as a calendar to record times of divergence and so forth.[7] Combining the information from radioactive studies with the information from neutral molecular studies, one had a very powerful tool to study the past and to come up with new findings.

None more than in the study of human evolution. Virtually overnight it was seen that the human line broke off from the great apes a lot more recently than anyone had hitherto dreamed, only about five or six million years ago. Moreover, we humans are more closely related to chimpanzees than chimpanzees are to gorillas. And so the story goes. Most recently, another molecular technique has come on board—ancient DNA.[8] It turns out that the DNA molecule, the molecule that is the molecular gene carrying the information needed to make an organism, is a lot more stable than people had realized. Stunningly, it has proven possible to sequence the genome (the set of genes) of Neanderthals, giving an answer to a question that had intrigued people since before Darwin even. Are we modern humans related to the Neanderthals, in the sense that there were sexual relations between us? Or were they just too ugly that even the most depraved of *Homo sapiens* drew the line there? Apparently, although not to a great extent, there was significant sexual activity between the groups, with the result that we now carry around 5 percent Neanderthal DNA. Or rather, we of European ancestry now carry around 5 percent Neanderthal DNA. It is not to be found in Asia or in Southern (sub-Sahara) Africa. Who are the cavemen now?

Causes

The talk of Neanderthal DNA starts to push us toward causes, obviously a matter of great interest and equally obviously a matter of much speculation.[9] It seems generally agreed that the big event five million or so years ago was the drying up of the lands where the jungles and forests flourished that were our arboreal homes, like those of the other great apes. Our ancestors left the trees and moved out onto the plains. It wasn't an overnight business. *Ardepithecus ramidus* lived about a million years before Lucy and although upright was much better at climbing trees than we or she.[10] For a long time, hominins lived on the flats and in trees, perhaps using the latter for safety at night. Why there was the move to bipedalism is controversial, and there may not be one simple answer. Plausible suggestions include the benefits of standing upright and thus more easily viewing the surrounding landscape. If the other denizens of the plains or savannahs include carnivores, then knowing about them before they know about us has clear advantages. Also, standing upright minimizes the heat from the sun, no small matter in Africa. And possibly a big advantage was that although we are never going to be able to run that fast absolutely, being bipedal rather than knuckle walkers like the apes means that we could cover large tracts more readily without tiring, thus traveling overall more quickly.

What about the brains?[11] General agreement is that the secret here is meat. Without good chunks of high-quality protein, available on a regular basis, there are not going to be big brains. Fussy graduate students would be in trouble. Even if one can live on tofu and lettuce, the natural equivalents of health-food stores were not then on offer. Where then does the protein come from? Obviously, the bodies of other vertebrates. Our ancestors were able to access these bodies and eat them. Note that there was nothing unnatural about this. The great apes supposedly stick to plant foods. I grew up in a family where there was a steady influx of books extolling the virtues of vegetarianism and about how meat eating is unnatural. I am glad to say that the effects of these books were not long-lasting. After a week of fava beans and flatulence, my father would break down at the thought of a pork pie—part of his evolutionary heritage, because claims about the naturalness of vegetarianism were false. The great apes will eat meat—monkeys and the like—if they can get it. They relish it and pity the baby baboon that crosses their path.

How to get the meat? Comparatively, although one would not care to go one-on-one with a silverback gorilla, we hominids (humans and great apes) have never been that strong. At first we were the jackals of the primate world, waiting until carnivores had made their kill and eaten to their desire. Then we could move in and feast. Probably then there was a fairly familiar—and note the word "familiar"—feedback process. There was a selective premium on having the smarts to get bigger chunks than your neighbors, and if cunning and skill were needed, then so be it. Especially if cunning and skill came from having (in Dawkins's language) a bigger and better on-board computer. So those who did better tended to have offspring more talented than before, and they needed as much if not more meat, and a progression (in a relative sense) was started. Notice though that this could not be something in isolation. One hominin and one lion is not an equal contest. Two hominins and one lion probably isn't either. But two hominins working together and one lion starts to even things out considerably. In other words, there is a selective premium, not just on cheating lions but on working with your conspecifics. Sociality is at a premium. This would lead to anatomical changes and not just brains. If you are working together and you have deadly weapons—as Americans with their gun fetishes show only too clearly—you are likely to hurt each other. So there is a premium on cutting down on aggressive hormones and dangerous physical features like large canines. And that, of course, makes cooperation even more important. At the same time, with larger brains there is going to be more need of child care and more ability to offer it.

At this point, a particularly important biological principle becomes relevant. Evolutionists distinguish between what they call K-selection and what they call r-selection. In the case of the former, organisms have few offspring but they care for them. In the case of the latter, organisms have many offspring but let them fend for themselves. Elephants and herring. Neither strategy is good in itself, but generally K-selection prevails where conditions are stable and r-selection where unstable. Especially with the latter, you can see how organisms are ready to take advantage in a big way if things are going well. In the case of humans, clearly we were under K-selection pressures and having few offspring for which we cared. Note that, with the coming of intelligence, hominins were increasingly able to control their own environment. Moving into a cave during winter is a good strategy for avoiding the perils of ice and snow. Note also that relatedly, hominins were able to extend their range from Africa out to the rest of the world. General opinion is that this did not come all at once but in waves—the "out of Africa" hypothesis.[12] Populations were always in the thousands but sometimes they were quite small—"bottlenecks." This is thought a reason why *Homo sapiens* comparatively does not have huge genetic variation, and also why you are going to get separate groups, perhaps even subsets, like *Homo sapiens neaderthalensis* and *Homo sapiens sapiens* (us).

Without yet digging into mind and thinking, the really interesting thing about human evolution is that, as I said earlier, it is so familiar in so many ways. For instance, in the past fifty years or so we have learned a huge amount about the evolution of social behavior, at all levels in the animal kingdom, and about how important it is and how natural selection has been so intimately involved all of the time. To take but one example, the American evolutionist David Reznick has ongoing studies of the evolution of guppies, the little fish found (in his case) on the Caribbean island of Trinidad.[13] He is particularly interested in such issues as life cycles and why some fish develop very rapidly and come early to maturity and other fish are more leisurely in their development and sexual behavior.[14] He ties these sorts of issues into matters to do with predation, habitat, and a host of related features. Organisms at both the social and the physical level do what they can to maximize their reproductive abilities and successes. And these issues involve both natural and sexual selection. If large males are dominating the group, is your best bet to take them on in combat or to take a strategy with a name too vulgar for this book but which involves waiting for opportunities when the leaders are not looking? And how do you do this? If you are top fish, then you can take your time over sex. If you are not, you had better develop adaptations for very efficient quickies.

> If the risk of being killed by a predator is high, then natural selection will favor those individuals that devote a larger slice of the pie to reproduction by beginning to have babies when they are younger and making more babies. . . . Conversely, if predators are scarce or absent and the risk of dying is low,

> then this long life expectancy shifts the balance in how best to invest resources. The theory then predicts that natural selection will favor those individuals who devoted a smaller slice of pie to reproduction and a bigger slice of pie to their own maintenance. The theory also predicts that they will produce fewer babies and devote more resources to each of them.[15]

The human story is different, of course. Every story is different. But it is the same sort of story. Think of the frenetic coupling that went on during the First World War when soldiers came home on leave—"hasty weddings" was a phrase that entered the English language. Natural selection (and sexual selection) grabbing the opportunities and molding organisms accordingly. No need for principles unknown or for throwing one's hands up in defeat.

Purpose?

Return again to the question of purpose and ends in human evolutionary history. I have already expressed my own skepticism on this score, and I am not sure that now with our fleshed-out knowledge of human evolution that I want to change my opinion. The temptation always is to think that with the growth to higher brains we do have progress. The purpose of evolution was to produce beings with large brains, namely, *Homo sapiens*. Actually, of all people, Gould seems to have been tempted this way once, especially if (as we did above) you tie in the process of brain production with K-selection and r-selection. In his major book *Ontogeny and Phylogeny*, the work that many (myself included) think is his greatest contribution to scholarship, he writes that he has "tried to link K selection to what we generally regard as 'progressive' in evolution, while suggesting that r selection generally serves as a brake upon such evolutionary change. I regard human evolution as a strong confirmation of these views."[16]

Which does rather raise the question of why Gould suddenly became so ardent against biological progress. My suspicion (backed by discussions with him) is that it was less epistemological and more political and moral, thinking such progress leads readily to racism with white Europeans at the top and Jews and blacks down the tree. His work *The Mismeasure of Man* supports this explanation.[17]

I can say only as I have said before that human evolution isn't really progressive, at least not in any easy way, and the growth of large brains doesn't alter things much. Sepkoski with his earthy comment about the virtues of being dumb and the center of the herd puts an end to that kind of thinking. Darwin thought the Scots K-selected and the Irish r-selected and a more general theory of religion suggests that Catholicism promotes r-selection because of the variability of life where it flourishes whereas Protestantism conversely promotes K-selection.[18] But neither Irish nor Scots look like they are going extinct, nor do Catholics and Protestants. As always with selection, it depends on the circumstances. In any case, as pointed out, Neanderthals had bigger brains than we. Is evolution now in decline? And what if rather, as in the 1985 novel *Galapagos* by Kurt Vonnegut, a disease had wiped out *Homo sapiens*, leaving only isolated *Homo floresiensis*? Would that disprove progress, for all that the hobbit seems to have been very well adapted to its way of life? Vonnegut, for what it is worth, would have denied this. His survivors turn into seals, and he thinks this a very good thing. Big brains and intelligence are not good adaptations.

Note also that there is good reason to think that human evolution is not yet over and that progress here is going to be relative. One of the most recent and important new adaptive moves was toward lactose tolerance.[19] Many adult humans cannot digest milk. Then, with the domestication of cattle, a new and valuable foodstuff became available, and natural selection acted accordingly. Not in every case, of course—the latest hypothesis about Charles Darwin's long and unexplained illness is that he suffered for this reason.[20] But while it was very unpleasant for him, compounded by his wife's providing meals with huge amounts of cream, lactose tolerance as such is neither good nor bad. In a world without cows, who needs it? It is all relative. From the viewpoint of biology—at least, from the viewpoint of modern Darwinian biology—there is no more progress in human evolution than there is in evolution as a whole. And that, for some of us, is a rather comforting note on which to end this chapter. Whatever purposes we are about to find for human beings, they must be such as acknowledge our oneness with the rest of nature. Value for one is value for all. My dogs approve of that conclusion and so do I.

CHAPTER TEN

Mind

Darwin Rejected

The writer Thomas Hardy was raised a good Christian, a member of the established church. Then he read *The Origin of Species* and it all came crashing down. His poem "Hap," written in 1866, tells it all. It is not just that God does not exist but that with his going, we lose all meaning to life. There is no purpose.

> If but some vengeful god would call to me
> From up the sky, and laugh: "Thou suffering thing,
> Know that thy sorrow is my ecstasy,
> That thy love's loss is my hate's profiting!"
>
> Then would I bear it, clench myself, and die,
> Steeled by the sense of ire unmerited;
> Half-eased in that a Powerfuller than I
> Had willed and meted me the tears I shed.
>
> But not so. How arrives it joy lies slain,
> And why unblooms the best hope ever sown?
> —Crass Casualty obstructs the sun and rain,
> And dicing Time for gladness casts a moan. . . .
> These purblind Doomsters had as readily strown
> Blisses about my pilgrimage as pain.

As we move now to crucial issues about mind and meaning, about knowledge and morality, let us ask first about how the philosophers handled all of this. Not just the nonexistence of God—agnosticism or atheism pretty much became the norm in the profession (as is true today)—but the lack of meaning. The American pragmatists rode with things pretty well. Whether this was part of the general, late-nineteenth-century American vigor and rise to prominence and power—perhaps the technological search for what works rather than the disinterested scientific search for absolute truth—they found the challenge of Darwinism stimulating and thought provoking.[1] For someone like William James, the struggle for existence and natural selection translated readily into a theory of knowledge—ideas fight it out just as organisms fight it out.[2] No more, but certainly no less.

The British had a lot more trouble. Virtually to a person, they turned against Darwinism, thinking it bad science and irrelevant to philosophy.[3] The Cambridge philosopher Henry Sidgwick set the tone. In a very early issue of what was to become (and still is) the distinguished journal *Mind*, he launched a strong attack on Darwin and Spencer on ethics.[4] His influence left its mark on his students, notably G. E. Moore and Bertrand Russell. Moore, in his famous *Principia Ethica* (1903), made Spencer a prime example of those who commit what Moore called the "naturalistic fallacy," a violation of proper thinking akin to a breaking of Hume's law of is/ought, that you are not to move from statements of fact to statements of obligation. Little wonder then that "evolution could hardly have been supposed to have any important bearing upon philosophy."[5] Russell, even more so, was hostile to Darwinism, belittling pragmatism as a "power" philosophy and narrowly defining the true scope of inquiry so that an empirical science like Darwinism almost by definition could have no role.[6] "What biology has rendered probable is that the diverse species arose by adaptation from a less differentiated ancestry. This fact is in itself exceedingly interesting, but it is not the kind of fact from which philosophical consequences follow."[7]

The Austrian import Ludwig Wittgenstein outdid them all. Even in midcentury he was saying, "I have often thought that Darwin was wrong: his theory doesn't account for all this variety of species. It hasn't the necessary multiplicity."[8] Not that this matters too much, because, in his all-influential *Tractatus* some thirty years earlier, Wittgenstein had given Darwinism a firm heave-ho. "Darwin's theory has no more to do with philosophy than any other hypothesis in natural science."[9]

One asks why there was this opposition. It is true that at the beginning of the twentieth century, Darwinian theory as a functioning professional science was hardly well established, although already there was some very good work by people like the Oxford insect specialist E. B. Poulton and the London-based theoretician and experimentalist W.R.F. Weldon.[10] One senses, however, an almost willful ignorance of, a determination not to seek out, quality science. One looks for deeper reasons. For some, perhaps, the answers are obvious. As a member of the German-led armed forces in the Great War, Wittgenstein would surely have been exposed to some of the horrific Social Darwinian thinking of the German generals.[11] In the case of the British, casting things in terms of our trichotomy and giving the philosophers credit for being swayed by philosophical ideas, one senses that the Kantian/Darwinian option—naturalistic and deeply developmental—was simply a nonstarter. Even if there is no god—perhaps because there is no god—these British philosophers sought stability and eternal truths. In short, they looked for the philosophy that imbued British middle- and upper-class education at the end of the Victorian era, as indeed it did for me even in the middle of the twentieth century. I speak of Platonism. This philosophy fit Moore with his belief that goodness resides in some ethereal world of nonnatural properties. He did not conceal this, writing to an acquaintance, "I am pleased to believe that this is the most Platonic system of modern times."[12] Russell, first and foremost a mathematician, was even more explicit in his search for absolute truths. Autobiographically, he wrote:

> I came to think of mathematics, not primarily as a tool for understanding and manipulating the sensible world, but as an abstract edifice subsisting in a Platonic heaven and only reaching the world of sense in an impure and degraded form. My general outlook, in the early years of this century, was profoundly ascetic. I disliked the real world and sought refuge in a timeless world, without change or decay or the will-o'-the-wisp of progress.[13]

Russell himself later in life repudiated much of this way of thinking, although (unlike his coauthor Whitehead) he seems not to have given up on his distaste for evolutionary thinking, especially as applied to the philosophical world. The die was set. This way of doing philosophy, "analytic" philosophy, swamped—as it still swamps—most of the universities of the English-speaking world. There is probably no one answer to why this should have been and continues to be so. As much as anything, the lack of interest in Darwinism until well into the second half of the century was a function of the intense interest in the physical sciences, something reinforced by the influx of Continental thinkers fleeing the Third Reich. There was occasional acknowledgment of the relevance of Darwin—the influential midcentury American philosopher W.V.O. Quine, who, incidentally, wrote his doctoral dissertation under Whitehead, noted the place of evolution in our thinking about the world, as did his Harvard colleague John Rawls in his massive *Theory of Justice*—but generally bringing Darwinism into the discussion was the philosophical equivalent of making a bad smell at the vicarage tea party.[14] The pragmatists, apart from some interest in Peirce on semiotics, were conspicuous by their absence from curricula. It is true that Dewey was interested in education, something with an academic status below even sociology, but still. Can you imagine a department of philosophy in France without Descartes, or Germany without Kant, or Britain without Locke, Berkeley, and Hume?

It continues today. A year or two back, the distinguished philosopher of mind, Jerry Fodor, coauthored with cognitive scientist Massimo Piattelli-Palmarini a book with the title *What Darwin Got Wrong*,[15] which just about tells you everything. Again, one looks for deeper meanings—a yearning for the now-lost comfort and security of the maternal breast—since it is clear that no one is being swayed by knowledge of professional Darwinian studies. One searches in vain for analysis of the earlier-mentioned results found and interpretations made by Peter and Rosemary Grant after their near-half-century study of the finches of the Galapagos, or the studies of Jerry Coyne about speciation, David Reznick's work on guppies, or detailed and informed thinking about the findings of paleoanthropology, Donald Johanson's discovery of Lucy, *Australopithecus afarensis*, for instance.[16] When mention is made of recent work, for example, Sean Carroll's stunning evolutionary development ("evo-devo") findings about homologies at the molecular level showing that the processes of growth are shared by fruit flies and humans, it is egregiously misinterpreted as replacing Darwinism rather than complementing it—"the huge variety of extant and fossil life forms . . . is not only fully compatible with the high conservation of genes, but explained by it."[17] Oh, and when it comes to Darwinism and philosophy, "Quine was too subtle a philosopher to be fully satisfied by this explanation. He recognized that there was a circularity it in."[18]

What is interesting and pertinent in the context of our discussion is that, as the twentieth century ended, in the philosophical world—due in major fact to a renewed interest in his approach to ethics—it was Aristotle's beacon that was shining most brightly. Today, so-called virtue ethics, owing much to Aristotle's *Nichomachean Ethics*—as opposed to the deontological ethics of Kant and the consequentialist ethics of the utilitarians like John Stuart Mill—is virtually the standard position against which all others are to be judged.[19] In tandem, the critiques of Darwinism are often set in an Aristotelian context. In a celebrated recent book, *Mind and Cosmos*, the New York philosopher Thomas Nagel divides his attentions equally between belittling Darwinism and promoting an Aristotelian perspective. He writes that "as it is usually presented, the current orthodoxy about the cosmic order is the product of governing assumptions that are unsupported, and that it flies in the face of common sense."[20] Continuing: "It is prima facie highly implausible that life as we know it is the result of a sequence of physical accidents together with the mechanism of natural selection." Asking rhetorically: "In the available geological time since the first life forms appeared on earth, what is the likelihood that, as a result of physical accident, a sequence of viable genetic mutations should have occurred that was sufficient to permit natural selection to produce the organisms that actually exist?"[21]

Nagel himself, although he cozies up to people like Behe, favors some kind of strong, secular Aristotelian position. He tells us that we are presented with two options: blind law (which presumably means Darwinism) or "there are natural teleological laws governing the development of organization over time, in addition to laws of the familiar kind governing the behavior of the elements." Nagel continues:

> This is a throwback to the Aristotelian conception of nature, banished from the scene at the birth of modern science. But I have been persuaded that the idea of teleological laws is coherent, and quite different from the intentions of a purposive being who produces the means to his ends by choice. In spite of the exclusion of teleology from contemporary science, it certainly shouldn't be ruled out a priori.[22]

As you will see, there is much in Nagel's thinking with which I sympathize. Yet overall, as you must now realize, I am just not sure that a secular Aristotelianism has staying power. Without some powerful central force—an unmoved mover or some such thing—I doubt you can get the end direction that is supposed. I don't want to belabor this point here. I am more interested in making the positive case for the third option, one that really is entirely secular and mechanistic, the kind of option to which Kant aspired and that Darwin made a reality. The main reason for introducing this background discussion is to point out that, as we move forward into more overt philosophical territory, do not presume that we are among friends. This is not to ask for mercy or an easy ride—I welcome the challenge—but it is to put things in context.

Sentience

Now let's take up the $64,000 question. What about the mind? What about sentience? One line of thought urges us to ignore it, or at least to put it on one side. Thomas Henry Huxley claimed we are all just automata,[23] with consciousness simply sitting on top of the brain, which latter is working in a purely mechanical method doing everything that is necessary. This is a version of "epiphenomenalism," more informally characterized as the "whistle on the locomotive" position. We have to remember that Huxley had a somewhat ambivalent attitude toward natural selection, so that consciousness having no adaptive function would be no great loss or worry to him. Most people don't find this line of thought very convincing. In the words of William James, "It is to my mind quite inconceivable that consciousness should have *nothing to do* with a business which it so faithfully attends. And the question, 'What has it to do?' is one which psychology has no right to 'surmount,' for it is her plain duty to consider it."[24]

Considered in this light, we shouldn't be too scared of sentience. It really doesn't seem to be something that is going to wreck evolutionary theorizing.[25] Let us agree that Hume and the British had things right and Descartes and the French had things wrong. Animals not only have feelings but they have sentience, from rudiments to fairly well-developed levels. At least, they give evidence of this, for instance, when dealing with causal situations. "It seems evident, that animals as well as men learn many things from experience, and infer, that the same events will always follow from the same causes."[26] Hence, "they become acquainted with the more obvious properties of external objects, and gradually, from their birth, treasure up a knowledge of the nature of fire, water, earth, stones, heights, depths, &c., and of the effects which result from their operation."[27] Going a good way down this path makes good sense. Obviously, sentience in its broadest sense is a very powerful adaptation, but it isn't something peculiar, in the sense of demanding totally new principles of understanding. Modes of selection, K-selection and r-selection, for example, are applicable. More than that, looking at things from outside in a broad sense, the rise of sentience does make good biological sense, for all that the bigger the brain, the greater the need of protein.

Let us agree also that the ability to use tools is a good indicator of the rise of intelligence and ever greater use of the mind. In looking at human evolution, there are no big surprises. As the brain got bigger, so tool use started and became ever-more sophisticated. Our genus, *Homo*—bigger brained (650cc initially compared to Lucy's 450cc)—emerged rather less than two million years ago.[28] They made ever-better tools and began using fire, although there is discussion about the exact date when the latter was truly brought under our control as opposed to used on fortuitous occasions. Simultaneously, the ability to talk was starting to emerge in a big way. Although there is still controversy, no one today wants to deny that biology plays a big part in language acquisition and use. In the 1950s, Noam Chomsky showed convincingly that language is not just something purely cultural but that all languages, from Japanese to English, share certain innate deep structures—a kind of biological ground plan on which everything is based.[29] As it happens, Chomsky is not particularly keen on Darwinian selection, but his students and followers, like Steven Pinker, have shown in detail how the innate hypothesis does lend itself to Darwinian understanding.[30] The great adaptive significance of language goes without saying.

The flip side to our greater grasp of what we might call the software of language ability is the hardware of language ability, something else where knowledge has grown greatly.[31] There is some fossil evidence of the actual physical apparatus needed to speak. The dropping of the larynx, for instance, is something that distinguishes us from the apes. Also, there is evidence of parts of the brain that are used in speech developing about two million years ago, specifically, Broca's area and Wernicke's area. Relatedly, we are now starting to identify some key genes involved in language acquisition and use, and showing that they are under the force of natural selection. Again, there is discussion and controversy. For a while, a popular hypothesis was that the Neanderthals, like the great apes, seem to lack the ability to talk properly.[32] This suggestion has now come crashing down with the discovery of a Neanderthal hyoid, a bone that is found in the throat and has essentially the function of enabling speech.[33] No one is suggesting that there was a Neanderthal Shakespeare—in fact, general opinion (based on the rise and development of culture) is that it was not until about 50,000 years ago that true speech emerged, and that it was probably some kind of primitive click language, found almost exclusively in Africa. But the Neanderthals were not dumb brutes. So this rather gives the lie to the underlying presuppositions of the novel *The Inheritors* (1955), by William Golding (better known for his *Lord of the Flies*), who saw the Neanderthals as innocent, inarticulate folk, who were wiped out by the brutal humans. It may indeed be that, when they were not copulating with them, humans did their bit in killing Neanderthals. Overall, however, what we seem to have had is in many respects a very familiar evolutionary selection-driven pattern, with branching, competition, relative improvements, and other similar occurrences and phenomena.

The Collapse of Darwinism?

My position about the evolution and use of mind is deeply rooted in Darwinism—I see mind as another adaptation, like hadrosaur honking tubes. It is also deeply Aristotelian in running together the purely organic and the mental. "Things that are done for the sake of something include whatever may be done as a result of thought or of nature."[34] I don't see that acknowledging this threatens the Darwinian stance at all; although, expectedly, in the light of what was said at the beginning of the chapter, a lot of people don't much care for (or are indifferent to) that elision. Nagel uses the special status of mind, one major arrow in his quiver, as part of his all-out attack on the theory. "Biology may tell us about perceptual and motivational starting points, but in its present state has little bearing on the thinking process by which these starting points are transcended."[35] He argues that the problem of mind shows that mechanism, meaning—especially meaning—evolution through natural selection is not only unsupported and implausible but also wrongheaded. Or at the least, gravely deficient. "An account of their biological evolution must explain the appearance of conscious organisms as such."[36] Continuing: "Since a purely materialistic explanation cannot do this, the materialist version of evolutionary theory cannot be the whole truth."

Nagel calls himself a "neutral monist,"[37] presumably seeing mind and body as emanating from the same stuff (whatever that might be), writing that "the failure of psychophysical reductionism"—explanations of mind in terms of the material—suggests that "principles of a different kind" are "at work in the history of nature, principles of the growth of order that are in their logical form teleological rather than mechanistic."[38] He keeps going in high gear, speaking warmly of Aristotle and (later) of Bergson, telling us that from life's first appearance to the arrival of humans "the process seems to be one of the universe waking up."[39] And concludes by chiding mechanists for ignoring values: "Value enters the world with life, and the capacity to recognize and be influenced by value in its larger extension appears with higher forms of life. Therefore, the historical explanation of life must include an explanation of value, just as it must include an explanation of consciousness."[40]

As it haunts Nagel, the problem of mind has always haunted me. Yet is it a problem here, for the evolutionist, in quite the sense Nagel implies? He runs the mind-body problem together with the as-yet-incomplete search for a naturalistic explanation of life itself and of its origination, but that is a mistake. No working biologist today feels the need to suppose a kind of special neo-Aristotelian life force, something along the lines of the élan vital—thinking that a live cow has it and a dead cow does not. While the origination problem is as yet unsolved, major work has been done in that direction, and no one thinks it needs the kind of conceptual or ontological jump that the move from body to mind seems to demand.[41] The origin-of-life problem is a red herring. We know what a solution would look like. With respect to the mind-body problem, I am with Nagel in that I am not sure we know what a solution would look like.

That said, I suspect the most popular position today about what has been labeled the "hard problem"[42]—the arrival and nature of mind—is some form of "emergentism," thinking (with people like Samuel Alexander) that, so long as you have got things organized in the right way, mind somehow rises up from the material, like Brigadoon appearing out of the mist. The trouble here is really that of freeing yourself from Cartesian dualism, with its problems. The mind is essentially different from matter, and that means that you still have the powerful objection of Leibniz (in the *Monadology* of 1714) that if you have a machine, a brain that is a (mini) factory or a computer or some such thing, it is hard to see where consciousness comes into it all, or why if it is not there in the physical brain, it nevertheless emerges up from it. Very trendy recently has been the notion of "supervenience," where a change in one domain is always associated with a change in another domain. You get the brain buzzing in a certain way. You get the mind buzzling in harmony. Somehow the mind supervenes on the brain. Unfortunately, there remains the nasty suspicion that this move is simply defining the problem away. What is the relationship? "Is it a matter of causal dependence?" Or is it something nonnatural? "Perhaps, a matter of divine intervention or plan as Malebranche and Leibniz thought? Or a brute and in principle unexplainable relationship which we must accept 'with natural piety,' as some emergentists used to insist?"[43]

Possibly emergentists have got the problem backward. Instead of starting with the material, the brain, and working to the mind, we should start with the mind and work back to the brain, to the material.[44] This starts to edge up to a form of monism, and I am certainly not going to turn from it because it is Nagel's position in some sense. The problem (a concern, that to be fair, is shared by Nagel) is that, too often, simple monism (meaning mind and matter are one substance) gives rise to crude panpsychism, where mind is functioning everywhere and (using the term in a strong sense) one has pervasive consciousness. Before long one has loonies like the Prince of Wales who talks to his plants—apparently they are his "friends." Actually, for once, the prince is in good company. The nineteenth-century experimental psychologist Gustav Fechner had similar views. I suspect most people react as I do. No one denies that plants have very sophisticated adaptations for sensing their environment and even changing it to their own ends.[45] But don't overdo it, else one will be wondering why they can't get into Harvard. Perhaps there are quotas.

I have noted that Sewall Wright, incidentally, brilliant scientist though he was, to the great embarrassment of his graduate students had yearnings in the direction of panpsychism. Whitehead inevitably went one step further. That mysterious word "prehensive" is about the way in which physical objects like electrons, no less than living beings, actually incorporate perceptions into themselves, so physical things are not material things but perceptions and the relations to other things. But, Wright or Whitehead notwithstanding, one doesn't have to agree that everything material is actively conscious—we humans are not a lot of the time—rather that in some sense the material is not just brute matter but a lot more than that. It is not that molecules are thinking. Rather that there is something about the individual molecule that gives rise to thinking. It is there in the individual molecule. You don't have to wait, as emergentism suggests, to get molecules put together before you get the whole new dimension that leads to thinking—although as a matter of fact, you are not going to get full-blooded thinking until you do get a lot of molecules put together. Like red paint getting redder and redder as you add more pigment, so consciousness becomes more and more aware as it is added to. It is a matter of increment, not innovation. Panpsychism in this more modified sense is another thing and nothing like as stupid as tradition has. Is this still "monism"? William James claimed to be a "panpsychic pluralist," meaning that he favored something along the lines of Leibniz's *Monadology*, with lots of separate mind-atoms; although later he claimed to be a "pluralistic monist," which rather confirms my feeling about James—contrary to Russell's characterization of the pragmatists, he is an incredibly lovable and sympathetic thinker, at times rather inclined to bafflingly foggy metaphysics.[46] Pluralism on one side—anything close to *Monadology* is nigh self-refuting (in a transcendentally magnificent way)—I don't see why one shouldn't speak of monism, in the sense that one is trying very hard not to rip apart mind and matter. It certainly was for Ernst Haeckel: "One highly important principle of my monism seems to me to be, that I regard all matter as ensouled, that is to say as endowed with feeling (pleasure and pain) and with motion, or, better, with the power of motion."[47] This is getting a bit close to royal family–type thinking, for my taste, but I take his point.

Thanks to modern physics, we know already that old materialist positions cannot be true, with the paradoxes about ultimate units being in some sense both particles and waves. Care is needed here. Lots of silly things have been said about modern physics and the mind, for instance, that quantum phenomena prove free will. That is nonsense. As Hume realized, if something happens randomly, that is not freedom but craziness. The point rather is that we now know that physical reality simply is not Cartesian *res extensa* and nothing more. Add to this positively spooky claims (Einstein's language) about such things as quantum entanglement where what happens in one part of the universe apparently is linked intimately with what happens in another part of the universe. Even if there is not a causal connection, there is an information connection.[48] While we certainly do not have a solution to the body-mind problem—and frankly talking about "panpsychism" may be little more than a fancy confession of ignorance—equally certainly we know that there is more there than meets the eye, and in the next century or two we might learn things that will surprise us a great deal.[49] Talk of "information connection" does start to push you to the view that sentience really is an aspect of the material. We have monism and, if not full-blooded panpsychic monism, some weaker form. Perhaps, ultimately, the Leibnizian objection about machines not thinking will prevail, but we might learn something significant and pertinent. Perhaps we shall be pointed to another all-powerful metaphor that will replace the machine metaphor without thrusting us back to the organic metaphor. We—or rather our descendants—shall see.

The all-important point to grasp is that, despite Nagel's naysaying, in many respects, the coming of consciousness, of sentience, has been handled remarkably well by the Darwinian evolutionist. It seems to have appeared and to have developed gradually, which is what one might expect. So long as you are not an epiphenomenalist, which applies to neither Nagel nor me, it seems to have obvious adaptive functions—information from without is received, it is processed, and then action is taken—again, what one might expect.[50] No one is taking value out of the equation. It seems amenable to more refined analysis, for instance, about the feedback between the needs of large brains (supporting sophisticated minds) and the availability of fuel to feed those brains and how the brains themselves (or minds) might contribute to the finding of such fuel. And so on and so forth. No one—no Darwinian—is going around tearing out hair because sentience is such an anomaly that it simply doesn't fit into the Darwinian scenario at all. The scientific theory takes the mind as a given, as it takes the brain as a given. The functioning, interpreted software—not just the written program, which seems to me more physical than mental—and the functioning hardware. With the monistic approach—and I am just introducing it in a friendly way without fully endorsing it—the issues seem less.

One's frustration with someone like Nagel comes because, in some ways, he seems sympathetic to the approach, even—despite some earlier disavowals[51]—to the point of panpsychism: "My guiding conviction is that mind is not just an afterthought or an accident or an add-on, but a basic aspect of nature."[52] Yet he flatly refuses to see that Darwinism is offering him pieces of philosophical candy! Others have been more grateful. One who did express his appreciation was the mathematician-philosopher William Kingdom Clifford:

> [W]e cannot suppose that so enormous a jump from one creature to another should have occurred at any point in the process of evolution as the introduction of a fact entirely different and absolutely separate from the physical fact. It is impossible for anybody to point out the particular place in the line of descent where that event can be supposed to have taken place. The only thing that we can come to, if we accept the doctrine of evolution at all, is that even in the very lowest organism, even in the Amoeba which swims about in our own

> blood, there is something or other, inconceivably simple to us, which is of the same nature with our own consciousness, although not of the same complexity.[53]

There is something really important being said here. The trouble with dualism or varieties of emergentism—theories that somehow make sentience separate from matter—is that when sentience comes on the scene, it is almost miraculous: "How it is that anything so remarkable as a state of consciousness comes about as the result of irritating nervous tissue, is just as unaccountable as the appearance of the Djin when Aladdin rubbed his lamp."[54] Now you have one thing, matter, and now you have something completely different, mind. How could something as strange as mind suddenly appear on the scene? I am not really surprised that someone like Robert Wright,[55] who calls himself an epiphenomenalist but who when mind turns up starts to sound much like an old-fashioned dualist, has problems to the extent that he starts talking again cryptically about "purpose"—meaning Purpose, something out of the normal course of nature. But pseudo-problems and untenable hypotheses like these are issues raised by the metaphysics of dualism/emergentism—supporters make mind different and then complain they cannot explain it—rather than by Darwinism. To the evolutionist, mind from nowhere just doesn't make sense, either as a question of general principle or empirical fact—oysters, ants, alligators, shrews, dogs and chimps, humans, philosophers. This is not a phylogeny, actual line of descent, but shows that sentience is not an absolute out or in sort of phenomenon. Never lose sight of this.

You might still ask for more. What really is mind and its relationship to the brain? Perhaps in the end it is legitimate to respond that, much as we might like an answer, it is not the job of the evolutionist to supply one. Take gravity. Gravity is very important to the evolutionist—the Darwinian evolutionist, that is. Why are there no cats as large as elephants? Because bodyweight goes up by volume and no elephant-sized cats could have the slender, supple legs of felines. The legs have to be tree-trunk-sized like those of elephants to carry the weight.[56] Gravity is the underlying principle here, but no one expects the Darwinian to explain the nature of gravity. Like the cookbook says, "First take your hare." Why then should the evolutionist be expected to explain the nature of consciousness? Surely it can be taken as a given, and the evolutionist can move on? It would be nice to explain consciousness as it would be nice to explain gravity. Perhaps explaining consciousness would give us new evolutionary understanding, as perhaps explaining gravity would give us new evolutionary understanding. But in science you never get everything you want, at least not at first. Leave the discussion at that.

Reasons and Causes

So we come to the heart of the matter. Nagel is absolutely right—mind does make things deeply, irreducibly teleological. Mind is the apotheosis of final cause, drenched in purpose.[57] It's all about values. Presumably, if one is someone like Nagel, inclined to some kind of monism, one runs with this, seeing mind and hence principles of teleological ordering pervasive in everything. Thus, metaphors like "waking up" make good sense. Although this is a little too Aristotelian for my taste—especially if you think in terms of the universe or the world itself waking up—I am not now going to raise a hard-line, epistemological objection. As we have seen, the move to the machine metaphor from the organic metaphor was not so much one of logic as of being able to do better and stronger science—more predictions and so forth. If the facts so dictate, as perhaps in optics, you can certainly go back to the earlier metaphor and take something of an instrumentalist attitude about prime movers and so forth. However, I am not sure that monism necessarily forces you toward an all-pervasive mind, a kind of world soul (which, at least in principle, could wake up), and at the present state of knowledge—remember, unlike Nagel I think Darwinism is a friend and not something that makes one eager to jettison the machine metaphor—one might still opt for a more cautious Kantian/Darwinian approach to things. Perhaps by talking of instrumentalism, you have already embraced this option.

Even going cautiously, purpose has a major role to play. Take, as an example, my daughter Emily—named after the poet (they share birthdays)—just turned thirty. She is a lawyer, a very junior public defender in Jacksonville in Florida. Her parents are inordinately proud, not just that she is a lawyer but that she is in a job serving others, society's truly down and out. Her parents are also very relieved that she is now making money to support herself! Ask now about how she came to be a lawyer. It was about as far from chance as it was possible for something to be. As she grew up, it was apparent to all that Emily was very vocal; she likes to talk. She is very social and bright in that sort of way. When it was a matter of putting together study groups, Emily was always a leader. She will not give up on something when she thinks it right and important. She has a concern for others in need and a rather brutal way of satisfying this. All through college she mentored a very handicapped student, and others were ordered to join in and help. In short, she had all of the qualities designed to drive her parents crazy and to make her what biologists call "preadapted" to be a lawyer. So she took the LSAT exam, went to law school, passed the finals and the Florida Bar Exam, and qualified. After several months of parentally supported volunteer work, she landed her job.

Now, let's have a look at what is going on here. Three major points stand out. First, qua evolution, something strange and unusual is happening. We don't just have nails being hammered into blocks of wood, or pheromones telling ants in a nest which direction they should go to find food. A lot of reasoning is going on here. As an undergraduate, Emily started to think about what she might do. She talked to friends and understanding adults. She may even have talked to her parents. She realized that there were some things she just wasn't able to do, like become a professional tennis player. She realized also that there were some things that she simply didn't want to do. She was never really attracted to being a teacher or a doctor. She saw that law was an option, and the more she thought about it, the more attractive it became. It's a worthwhile job; it can be very interesting; it pays reasonably well, notwithstanding that one incurs horrendous student debt; it has a reasonable status in society, for all that her father kept quoting Shakespeare ("First, we'll kill all the lawyers"). And, most important of all, it means you have to wear nice clothes. So these thoughts, desires, and intentions kicked in, and several years later the weekends found Emily at the mall looking for something to wear on Monday morning.

So we do have something strange, but is it that odd? Going back to Plato and the *Phaedo*, we find that, as a matter of course, quite naturally he brings us humans into the picture as entities who are going to be part and parcel of the purpose story. Take Socrates himself. At one level, he is sitting in prison awaiting the hemlock because of his physical nature—his bones and flesh and so forth. But that is hardly the *reason* why he is there. The bones and flesh would have had him out of prison and far away long ago "if they had been moved only by their own idea of what was best, and if I had not chosen the better and nobler part, instead of playing truant and running away, of enduring any punishment which the state inflicts." Continuing: "There is surely a strange confusion of causes and conditions in all this. It may be said, indeed, that without bones and muscles and the other parts of the body I cannot execute my purposes. But to say that I do as I do because of them, and that this is the way in which mind acts, and not from the choice of the best, is a very careless and idle mode of speaking."[58]

We are giving things a Darwinian interpretation, but we have seen that somehow the mind game seems to fit into the evolutionary picture, so we know already that it cannot be that odd. It is really not like the way that Rudolf Otto described God, as "numinous," as unknowable, as the "Wholly Other."[59] For instance, just as if a nail is driven into the wood by a hammer, it cannot have been driven in by a staple gun, so if Emily was motivated by the idea that law would be interesting and fun, she cannot have been totally motivated by status and money. The usual laws of logic and so forth seem to apply. For me, a good analogy is with the square root of minus one: $i^2 = -1$. We all learned at school that although minus numbers can be square roots, they cannot themselves have square roots, because when you multiply a number—positive or negative—by itself, you get a positive number. And yet then, you find out that mathematicians do want to talk about the square roots of minus numbers, most famously *i* the square root of -1. At least they have the good grace to call them "imaginary numbers." What you also start to learn is that although imaginary numbers are very peculiar things with apparently no real-world referent, they are not that peculiar. You can add them and multiply them and so forth. Moreover, you can include them in equations where they seem to function perfectly normally. For instance, in the Euler identity: $e^{i\pi} + 1 = 0$. One goes on to learn that *i* has huge numbers of practical applications. In electrical engineering you can work out all sorts of complex problems using the square root of minus one.[60] I suggest that reasons function in much the same way as the square root of minus one. They are strange, but they follow logic and mathematics and so forth. It is better to have two good reasons rather than one, and one rather than none at all. And they are obviously very useful. Hominins, protohumans, our ancestors, used reasons very effectively to find food in quantities and of a quality that it would have been nigh impossible to get without thought and reasoning.

The second point is that, when I started out as a philosopher more than fifty years ago, many denied strongly that reasons (that is, reasons with outcomes we desire) can function unambiguously as causes—efficient causes, that is. It was argued that reasons get us into a whole new, noncausal ball game.[61] The classic text was Elizabeth Anscombe's Wittgenstein-influenced little book *Intention* (1957). (As a Roman Catholic convert, there was also expectedly a lot of Aristotle—end-directed values, natural rather than imposed.) It is true that intentions are (or can be) about the future, but they are not causal, like predictions. "What distinguishes actions which are intentional from those which are not? The answer that I shall suggest is that they are the actions to which a certain sense of the question 'Why?' is given application; the sense is of course that in which the answer, if positive, gives a reason for acting."[62] This means that we are not so much into the business of empirical justification, as we would be with causes, but more into evaluation. Likening intentions to commands, Anscombe writes that "there is a difference between the types of ground which we call an order, and an estimate of the future, sound. The reasons justifying an order are not ones suggesting what is probable, or likely to happen, but e.g. ones suggesting what it would be good to make happen with a view to an objective, or with a view to a sound objective. In this regard, commands and expressions of intention are similar."[63]

What does this all mean? There is a strong, implicit message that intentional beings escape the forces of nature. We may be in the world of final causes. We are outside the world of efficient causes. But we can't so easily escape the forces of nature and we don't.[64] Of course, when we first start into discussion about reasons, Anscombe is right that we are usually less concerned with efficient causation and more with some kind of understanding and evaluation.[65] Final-cause thinking. Barbarossa. Why on earth did Hitler invade Russia on June 22, 1941? Did he learn nothing from what happened to Napoleon? Clearly not, for he left things so late in the season that, thanks to the autumn weather and oncoming cold, his troops were bound to get bogged down. Rather, Hitler was encouraged by his success against France in the previous year and even more by his sense of destiny, that Fate or the Immanent Will had picked him to lead his nation. All else paled in comparison. So we can understand why Hitler reasoned as he did.[66] Then, obviously, we who are looking at episodes like this start to move on. Often people have very good reasons for doing things but just don't do them. It is here that reasons, along with general temperament and external forces and so forth, get kicked into causes (or not), and we do things (or not). "We cannot explain why someone did what he did simply by saying the particular action appealed to him; we must indicate what it was about the action that appealed. Whenever someone does something for a reason, therefore, he can be characterized as (a) having some sort of pro attitude toward actions of a certain kind, and (b) believing (or knowing, perceiving, noticing, remembering) that his action is of that kind."[67] Why did Hitler follow his instincts rather than listen to history? Because at some level, his reasons convinced and drove him to action. He thought (a) he was guided by destiny and he believed (b) that marching into Russia was a specific instance of being so guided.

Take Anscombe's own discussion and use it against her in a standard natural selection situation. She admits that animals can have intentions. "Intention appears to be something that we can express, but which brutes (which e.g. do not give orders) can *have*, though lacking any distinct expression of intention."[68] Consider lions. Apparently, the females do the hunting. The males wait for the catch and, using their superior strength to help themselves, move in. A female lion goes (intentionally) to the top of the gully, hides in a thicket, waiting to jump out when the buck gets close. A second lion goes (intentionally) down the gully and waits, and then when the frightened buck comes racing down, grabs it and kills it. Intentions, reasons or "reasons," if you like—Why did the lion hide in the thicket? To scare the buck—but overall about as causal as you could possibly imagine. It is not a question of causes or reasons but causes and reasons. Efficient and final causes. "The origin of action—its efficient not its final cause—is choice, and that of choice is desire and reasoning with a view to an end."[69] The lions' behavior was causally adaptive—they did what they did in order to survive and reproduce—as does the totally unconscious Venus flytrap when it snaps shut on some unfortunate insect that has wandered within its orb. The fact that the lion scenario is all about values—the lions want to catch the buck, because for them this is a good thing—is far from being a problem and taking us from a causal analysis, precisely what we expect and demand. After all, the plant wants to trap the fly—from the plant's viewpoint, it is a very good thing.

All of this is precisely what we would expect. Overall, whether dictator or lion, we are animals and part of the real world. On the one hand, James was right and Huxley was wrong. Evolution through natural selection simply doesn't produce and cherish expensive items like functioning brains if they are not going to make a difference in the real world, the world of efficient causation. On the other hand, if our minds (using our brains) didn't function as superb causal machines, we would have gone extinct long ago. The nail is deeply embedded in the block of wood. Why? Because I hammered it in. Emily is a lawyer. Why? Because several years ago she got the idea of being a lawyer and, having researched things, set herself on a track that has just ended in the public defender's office in Jacksonville. If you say the nail is in the wood because Lizzie my wife took a stick of butter to it, you would be wrong. If you said Emily is a lawyer because she had a vision and Jesus told her to be a lawyer, you would be wrong. Jesus was with Shakespeare on this one. "And he said, 'Woe unto you also, ye lawyers! for ye lade men with burdens grievous to be borne, and ye yourselves touch not the burdens with one of your fingers'" (Luke 11:46). Of course, reasons are not causes exactly, like hammering in nails, but we know that already. Equally, of course, one reason is probably not the only cause. Emily is a lawyer because there was a loan system that helped her to pay the fees at law school. None of this is at all that odd. Have you ever tried hammering a nail into a piece of wood with the hammer in the one hand and the wood in the other? You need a bench or a support to do the job, and this is surely part of the relevant causal network.

I appreciate that someone like Anscombe in a neo-Aristotelian world could never find adequate the stripped-down Kantian-

Darwinian analysis I seek. So, let me note that, while critical, I accept entirely positive aspects of her thinking, such as the recognition of value—in a way, we are in complete harmony in seeing that forward-looking thinking demands more than simple reference to efficient causes. You need final-cause thinking. My aim is to deny the nigh-paradoxical claims for reasons, and to get away from a kind of up-on-a-pedestal view of them. I see purpose (or, rather, purposes) right through the living world; I argue that humans (and other sentient beings) are not different in being uniquely purposeful, and I argue that such beings are (like with the square root of minus one) in possession of a new tool that gives a way more powerful way of having and satisfying purposes. That means reasons and values, but it does not push out efficient causation. Emily took the LSAT exam in order to become a lawyer. She had her reasons! The point is that the reasons refer to the fact of becoming a lawyer. It turned out—by the time she had thought things through, taken a year off to travel, and so forth—this was an event some six or so years in the future. But if we know anything by now, we know that there is absolutely nothing tense-making here. It is not an either/or but not both situation. We are surely sufficiently with Aristotle to know that final causes pretty much demand efficient causes. The purpose of taking the exam (around 2008) was to become a lawyer (in 2014). There is nothing odd in the sense of little men in the future manipulating the strings of the present or anything like that. (Spinoza in the appendix to the first part of his *Ethics* made that point, in the context of an argument strongly criticizing Aristotelian final causes.) The missing-goal-object problem still applies. In the year she took off to travel between undergraduate college and law school, she went to Australia. It is quite plausible that she might have stayed there and ended up doing something entirely different, like becoming a sheep farmer. You get to ride horses. It is just that, because Emily's reasons included thoughts about the future, and that is what motivated her, we are dealing here with a purpose-oriented situation. We are talking about values. We are also talking about causes.

The third and final point is that, of course, things are not exactly the same as in the straight physical example. To give an example where there is a clear (temporal) gap between efficient cause and final cause, the child's testicles exist in order to reproduce in the future. The testicles are around because of a long line of testicle-possessing reproducers. Kant pointed that out and Darwin confirmed it. Emily is not the end point of a long line of successful lawyers. Although she would not have set out to become a lawyer if everyone she knew who had taken the LSAT exam had failed to become a lawyer. So what happened in the past is not irrelevant. Nevertheless, it is somewhat different. What about the big question raised by Nagel's attack on Darwinism? Plato, Aristotle, or Kant—what Nagel calls "intentional," "teleological," and "causal," respectively? What is the right overall analysis here? Obviously, within the system, as it were, one has conscious design, as demanded by Plato. One might also say that one has the kind of forward-looking plan or system that characterizes Aristotle's approach to final causes. Emily did think about things and plan ahead. However, with respect, we know all of that already, and unless you simply by fiat take things out of the natural order, you have to push a little further. You cannot simply be an Aristotelian like Nagel. Looking at those testicles, a cautious thinker (like myself) feels a Kantian analysis—heuristic, completely mechanical—is most appropriate because we think the testicles are just testicles, if we might so describe them, and the purpose thinking is imposed on the situation by us.

> The concept of a thing as in itself a natural end is therefore not a constitutive concept of the understanding or of reason, but it can still be a regulative concept for the reflecting power of judgment, for guiding research into objects of this kind and thinking over their highest ground in accordance with a remote analogy with our own causality in accordance with ends; not, of course, for the sake of knowledge of nature or of its original ground, but rather for the sake of the very same

> practical faculty of reason in us in analogy with which we consider the cause of that purposiveness.[70]

Although, in the case of genitalia, as I have conceded, there is something out there to which we are responding. Believe me, I am somewhat of an expert on these matters.

Likewise in the human case, we are structuring the situation but doing so in some sense responding to what is out there. If we look at Emily from the outside, as it were, then—as I have been stressing again and again—it all seems rather familiar. It is the square root of minus one all over again. She chats with her friends, she reads a book or two, she listens (or not) to her parents, and then she takes certain actions—sitting for the LSAT exam—and in the end she walks through the doors of the public defender's office in Jacksonville. Like any other healthy young animal, with obvious qualifications that will be raised in the next chapters about the dimensions of freedom brought on by culture and how we can hence, in some ways, escape from the brute, direct necessities of Darwinian existence, Emily is out there foraging for food and (undoubtedly before too long) reproduction. It is not as if Emily's mind—and this is not in any sense to knock it—has suddenly joined up with other minds in a kind of Hegelian sense now guiding the course of history. I am not saying there is no social progress—more on this in chapter 11—but that my daughter is a sophisticated and thus far rather successful organism, just like other organisms. She is one of Richard Dawkins's "survival machines." It is proper to think of her in terms of purpose. Her decision to become a lawyer has exactly the same kind of relationship to the future as the testicles have to producing babies.

Adaptability

We draw to an end of this part of the discussion, but there is a final point to be raised. We need to draw the distinction between being "adapted" and being "adaptable."[71] All organisms are adapted. Most, if not all, are adaptable. The concepts are related but not the same. For obvious reasons, one tends to think of K-selection for adaptability and r-selection for adaptation. Leaf-cutter ants are highly adapted to their surroundings. If foragers find suitable leaves, then they make pheromone trails back to their nest so that cutters can come out and do their business and carry the parts home. But they are not very adaptable. If a rain-storm wipes out the trail, the cutters away from home are lost and probably die. The nest can bear this cost because literally millions of workers are being produced. Mammals are adaptable with respect to outside temperature. They need to maintain a constant body heat—for humans it is 98.6 degrees Fahrenheit (37 degrees Celsius)—and if they get too hot they sweat, and if they get too cold they shiver, thus bringing them back to the original state. (Not always, obviously.) This is being adaptable, meaning that they can adjust things in order to keep the goal in view. Adaptability comes in many forms, and sometimes it is a one-off thing, where an animal might grow one way to adapt to conditions and might grow another way to adapt to different conditions. In all cases though, it is a matter of adjusting to stay adapted. This is often known, especially in philosophical circles, as being "goal directed" or "directively organized." A lot of earlier work was much influenced by successes of homing devices invented for weapons (like torpedoes) in World War II.[72]

The obvious thing about humans as organisms is that thanks to our thought processes, we are highly adaptable. Reasoning makes us very good at going after goals, thinking strategies through, and when barriers are raised, then reflecting, reassessing, and taking different directions to achieve the goal—or a suitable substitute. For Emily, becoming a lawyer required these sorts of decisions and reassessments along the way. When she got her offers of admission to law school, one was from a school in New Orleans. As you might imagine for someone with a sociable nature, the idea of three years in New Orleans was very attractive. The problem is that Louisiana law is unique in the United States, as it is based in part on French law. This would be no help in passing the Florida Bar Exam. It is one thing to spend three years in New Orleans. It is another thing to spend the rest of your life in Louisiana. So strategy decisions had to be made there. Then, after she had been admitted to the Florida Bar, there was much discussion about the right job, especially given her student loan. One good thing about working in a public defender's office is that, after a number of years, much of the loan is forgiven. And so forth, from beginning to end.

This matter of thought making us highly adaptable does not make us unique—it does not make other animals with levels of consciousness unique—but it does give us a powerful adaptation that we can and did use to advantage. It made us very efficient hunter-gatherers, for instance. It gave us the ability to be highly social. Developing sophisticated communication methods obviously helped here too. It led to tool use and then to tool improvement, as perhaps new prey in newly entered lands led to the need of different kinds of tools. One very much doubts, for instance, that the first bow and arrow sprang into existence fully formed and functioning. The same with making and using fire.

This discussion raises the ever-thorny question of free will. Adaptability means decisions, strategies—from within rather than from without. This is what freedom is all about. I am with the robust thinking of David Hume on its existence.[73] He thought it was just silly to claim that humans always act out of self-interest, and for all that there are those today who claim we have no free will, it seems just silly to say this seriously. Emily clearly had a choice about whether to go to law school in New Orleans or not. She was not just a falling rock, powerless to make a decision. If you take evolutionary biology seriously, it is hard to see why you would deny some kind of Humean analysis of free will, seeing (in a "compatibilist" manner) that free will does not deny that we are subject to causal laws.[74] Darwin himself saw this, if not entirely clearly. Sometimes (writing in private notebooks around the time in the late 1830s when he was discovering his theory), he denied that we can have free will because we are determined—"one doubts existence of free will every action determined by heredetary [*sic*] constitution, example of others or teaching of others."[75] But then common sense intervened and he admitted fully that animals can have free will. "With respect to free will, seeing a puppy playing cannot doubt that they have free will, if so all animals."[76] He even thought this might be true of oysters!

Certainly if the alternative ("libertarian") view holds, it is hard to see how humans fit so nicely into the Darwinian picture. Not surprisingly, Nagel leans this alternative way somewhat. Like Hume, I think that if reasons do not in some sense determine our actions, then we don't have freedom. As I have said, we have craziness. If reasons do determine our actions, then why not natural processes governed by law? A tide is pushing against the floodgates. Two forces, one prevails—the gates hold or the gates burst. Go to New Orleans or stay in Florida? Two forces, one prevails—she goes to New Orleans or she stays in Florida. Just as you can explain why the gates hold, so you can explain why she stayed in Florida. Given Emily's purpose—to get a decent job as a lawyer in Florida—staying in Florida for law school was the stronger force for making her decision. Leave it at that.

CHAPTER ELEVEN

Religion

Overall Purpose

Plato and Aristotle have had huge influences on religion, particularly the Christian religion. They had their say many times earlier in this book, so focus now on the attitude of the Darwinian. One temptation, in the tradition of Lucretius, is to dismiss it all as a dreadful mistake and move on. Richard Dawkins and his fellow New Atheists would be happy to do this. Let us take our time and ask especially how religion plays out with respect to questions about purpose. Obviously, thinking first about Christianity, we have answers both at the individual level and at the historical, big-picture level. Going first with the latter, to put things in context, God created humans to have what are essentially his children, to love and to cherish and in return to have them thank and adore and (not quite like human children in my experience) worship. The idea is that we should spend eternity in blissful joy with him. In many versions—the Augustinian version particularly—we humans rather spoiled things through our disobedience, but God in his boundless love sacrificed his son on the cross, and once again salvation is made possible. In both versions—Plan A, when we didn't sin and Plan B, when we did—purpose, teleology, final cause is the underlying theme throughout. God did not create just for laughs or because he was bored. He did it so that he could have creatures made in his own image to love and cherish. He wanted good for us, and he had plans that we would spend eternity with him. You cannot understand the Christian religion without this seizing on its end-directed vision. It is all a matter of values. God's values.

There are other non-Augustinian versions of Christianity,[1] Eastern Orthodoxy, for instance, not to mention variants in the West, like the Quakers and others more extreme, such as the Jehovah's Witnesses. These latter have little truck with traditional views, starting with a very iffy relationship with the Trinity (not for nothing are they called "*Jehovah*'s Witnesses"). Perhaps unfairly—partly for economy and partly because they were often not central to the workings out of science-and-religion relationships—I have rather ignored these other versions. So let me mention them now and stress just how far end-directed their theologies always are. There is no more eschatologically focused religion than that of the Jehovah's Witnesses, who are obsessed with the end of time and the subsequent 144,000 who are going up to heaven to rule with Jesus.

The same commitment to ends and values is true of other religions, both those sophisticated and those less so. Famously, Buddhism has no Creator God, but it too is purposeful throughout.[2] Central to Buddhism is the idea of reincarnation—that we have multiple lives in succession (*samsara*)—and that actions and thoughts in this life can have implications for the life that we will live next. There are levels of existence—down at the bottom is the hellish realm, *niraya*, and then up through the *petas* (ghostlike creatures), animals, humans, and to gods. Ultimately, the aim is to break out of this ongoing cycle of existences—one is released from suffering (*dukkha*)—and one achieves something called *nibbana* (also called "nirvana"). It is often thought that *nibbana* is a form of nonbeing, but this is not quite true, at least not quite true in all versions of Buddhism. As with other religions (notably Christianity), it is stressed that one is talking of the ineffable, the unspeakable, but then as with other religions (notably Christianity), people do go on to speak about it. *Nibbana* is endless and wholly radiant, the "further shore," the "island amidst the flood," the "cool cave of shelter" (no small thing given the Indian climate), the "highest bliss."[3] This doesn't sound altogether different from the Christian idea of heaven, except—what many would say makes for complete difference—there is no God there to share things with you. What is not different is that, in having the goal of *nibbana*, Buddhism is as clearly purpose-driven as is Christianity—or Islam with its seventy-two virgins and so forth. It is all a matter of values.

Going back before Christianity, and, indeed, most of today's major world systems, one finds various primitive or folk religions. Often these go under the generic name of "paganism," although the term is a little too generic if one simply means someone who falls outside the major religions.[4] It seems a little odd to link up Plato and Aristotle with people in what is now Norway who worshipped reindeer and did rather rude things under oak trees. One thing that did link many of these belief systems, not just with each other but with many of our philosophers, was a thoroughgoing commitment to a living earth, something of reverence and awe. Today, in Western society, there are revived elements of this kind of thinking. Some combine it with forms of Christianity. This is true of the Austrian polymath and clairvoyant Rudolf Steiner—founder of the Waldorf system of education—and of his followers (anthroposophists).[5] Thinking and working at the beginning of the last century, influenced in equal parts by Naturphilosophie and Eastern religions (especially through the theosophists like Madame Blavatsky), Steiner was totally committed to the Earth-an-organism view of nature. Through her intimate friendship with Steiner-follower Marjorie Spock (younger sister of the doctor), Rachel Carson—author of the very influential *Silent Spring* (1962)—showed his influence in her attack on those poisoning Mother Earth. Through his intimate friendship with Steiner-follower William Golding, James

Lovelock of the Gaia hypothesis showed his influence in his insistence that the earth is alive in some very real sense.[6]

Steiner was open about the pagan roots of much of what he believed. One doubts either Carson or Lovelock have ever thought quite in these terms, although Lovelock has an impish sense of humor and he might enjoy the label—he laughed when I wrote a book on Gaia with the subtitle "Science on a Pagan Planet." Whatever they are called, and if indeed Carson and Lovelock can be called genuinely religious—toward the end of her life Carson wrote (in a private letter) that "there is a great and mysterious force that we don't, and perhaps never can understand,"[7] and my impression of Lovelock is of a man who, in a totally nonprissy way, is deeply spiritual—we are looking at thinkers who live within purpose-laden worlds. There is the goal of a healthy, functioning planet. The same holds for others of today's nature worshippers who more openly identify with a non-Christian paganism. This is true of many "ecofeminists": "The physical rape of women by men in this culture is easily paralleled by our rapacious attitudes toward the Earth itself. She, too, is female,"[8] and "The planet, our mother, Grandmother Earth, is physical and therefore a spiritual, mental, and emotional being."[9] It is true also of those who set themselves up overtly as pagan or neopagan wizards and witches. Thus, Oberon Zell-Ravenheart (born Tim Zell): "It is a biological fact (not a theory, not an opinion) that ALL LIFE ON EARTH COMPRISES ONE SINGLE LIVING ORGANISM! Literally, we are *all* 'One.' "[10] And obviously uniting all of these people is a teleology no less thorough than that to be found in conventional religions. There is perhaps not the focus on distant ends—the hereafter—but our lives occur within a universe where nothing makes sense except we see ourselves as parts of a functioning whole. "The blue whale and the redwood tree are not the largest living organisms on Earth; the ENTIRE PLANETARY BIOSPHERE is."[11] Individual organisms are the cells of Terrabios (Zell-Ravenheart's name for Gaia). The deserts and the forests and the prairies and the coral reefs (the "biomes") are the organs. "ALL the components of a biome are essential to its proper functioning, and each biome is essential to the proper functioning of Terrabios."[12] There is purpose throughout.

Individual Purpose

Moving now down in scale, in all of these big-picture scenarios, the individual has to do his or her bit. It is often stressed, at least for Christians, that one should do things now because they are right, not with the intent of piling up brownie points in order to get into heaven. One should do good now, not because of hope of future reward but because God wants you to, or (as Augustine stressed) as a thankful response to God for his goodness. That said, the end of things does figure not just in the imaginations of many believers but also in the theology. Think of America's most famous sermon. "The God that holds you over the pit of hell, much as one holds a spider, or some loathsome insect, over the fire, abhors you, and is dreadfully provoked; his wrath towards you burns like fire; he looks upon you as worthy of nothing else, but to be cast into the fire; he is of purer eyes than to bear to have you in his sight; you are ten thousand times so abominable in his eyes as the most hateful venomous serpent is in ours."[13] If this isn't something telling you to watch your step because if you don't some pretty unpleasant times are on the way, then I don't know what is. There wouldn't have been one of Jonathan Edwards's congregation—it was preached in 1741—who would not have left church that morning thinking that they had better mend their ways because otherwise trouble lies ahead. All of them from that point on would have believed that the purpose of their lives was to keep from that fate.

Part of the trouble with Christianity is that of knowing just what will get you out of trouble. We have seen that this was a major concern of Augustine. Edwards's congregation apparently interrupted his sermon, crying out, "What shall I do to be saved?"

Edwards, a good Protestant, knew the answer—justification by faith. "For by grace are ye saved through faith; and that not of yourselves: it is the gift of God: Not of works, lest any man should boast" (Eph. 2:8–9). Other parts of the Bible contradict this flatly. "What doth it profit, my brethren, though a man say he hath faith, and have not works? can faith save him?" (James 2:14). There are deeply sincere Christians who think that works are everything. They have always been for me, raised as I was in the Religious Society of Friends. Remember the hungry and the strangers and the poor and the sick and the prisoners. "Inasmuch as ye have done it unto one of the least of these my brethren, ye have done it unto me" (Matt. 25:40). There follows a rather juicy passage about what happens to you if you don't do these things. Belief doesn't enter into the equation.

Why should you be good? For the Christian, morality and its foundations are all a matter of design, or rather Design. God set the rules and it is for us to follow them. "Who has a claim against me that I must pay? Everything under heaven belongs to me" (Job 41:11). There are well-known problems with morality being God's design, most prominently the *Euthyphro* problem. Is something good because God so willed it, or is God's will following that which is independently good? If one accepts the first half of the dilemma, then God seems somewhat capricious. If one accepts the second half, then God is not the ultimate authority. There are responses. Job responds to the first. God's answer to Job: "Where wast thou when I laid the foundations of the earth? declare, if thou hast understanding. Who hath laid the measures thereof, if thou knowest? or who hath stretched the line upon it?" (Job 38:4–5). Many would respond that the way God treats Job shows precisely the problem with this kind of approach. More acceptable in the eyes of many is a natural-law position, going back to Aquinas and thence to Aristotle (with a good shot of Cicero along the way), saying that the way God created the world and the way God expects us to act are at one.[14] Thus, for instance, heterosexual intercourse is in principle a good thing, because this is natural. Beating babies on the head is not a good thing, because it is unnatural. The point is that God sets the ends, the purposes, and expects us to follow them.

Under all versions of Christianity, your actions are judged and conclusions drawn about your future fate. Nothing is without purpose. It is the same in other religions. Buddhism is as end-directed as Christianity, although it does seem that if you mess up, you are going to descend a level or two, but then have the chance to climb back up again. The Four Holy (or Noble) Truths point the way. We start with self-examination, understanding the unsatisfactory nature of our lives—our greed and the like and how this leads to suffering. We must understand *dukkha*. Next we must understand the reasons for *dukkha* and our incomplete and selfish natures. We must grasp how it is that we can never feel full happiness in our lives. Third comes understanding how *dukkha* can be ended and *nibbana* achieved: "This, monks, is the holy truth of the cessation (nirodha) of dukkha: the utter cessation, without attachment, of that very craving, its renunciation, surrender, release, lack of pleasure in it."[15] Finally, we have what is known as the eightfold path of action—seeing reality as it is, renouncing desires, speaking truthfully, doing no harm, living in a wholesome way, trying to improve, making an effort to see oneself clearly, meditating. This is all part of karma, the actions taken that can affect the future lives. And obviously means that our lives are as full of purpose as anything to be found in Western religions.

The same is true even more obviously for all versions of nature worship. Rachel Carson was way too skilled a science writer to make explicit her beliefs in living earths—she knew that the established powers were going to be highly critical without giving them the opening of going after flaky notions like anthroposophy—but it is there underlying all of the exhortations. From her writings that she wanted read at her funeral service: "We come to perceive life as a force as tangible as any of the physical realities of the sea, a force strong and purposeful, as incapable of being crushed or diverted from its ends as the rising tide."[16] Continuing: "We have an uneasy sense of the communication of some universal truth that lies just beyond our grasp." Little wonder that: "The meaning haunts and ever eludes us, and in its very pursuit we approach the ultimate mystery of Life itself."[17] Very explicitly writing in this tradition—"For women making the connections between the masculinist ravaging of nature and the rape of women, Carson was a forerunner"—the ecofeminists openly urge environment-protecting strategies upon us. "With no sense of consequence in the scant knowledge of harmony, we gluttonously consume and misdirect scarce planetary resources."[18] And the same is true of the pagans. Our place here is "to act as the steward of the planetary ecology." This is our destined role: "Man's purpose in Terrabios, his responsibility, is to see that the whole organism functions at its highest potential and that none of its vital systems become disrupted or impaired."[19] These are the values we must embrace.

Religion as False

Not everyone is religious. There are those of us who are agnostic or atheistic. My concern now, however, is not with what I believe, or with what you believe, but how one analyzes religion on the Kantian-Darwinian perspective. New Atheists like Richard Dawkins think that Darwinian evolution disproves religious claims. He points out truly that children need to learn things quickly—to fear fire, for instance, and that natural selection has made us susceptible to conditioning. "Be fantastically gullible; believe everything you're told by your elders and betters."[20] Which, of course, is fine much of the time but open to invasion by parasites with their own interests in mind. It is very much the same sort of thing that happens with computers. Viruses invade with their own agendas, not necessarily in the interests of the hosts. Unfortunately, religion is right up there with the worst of the invaders. Dawkins thinks that humans are wide open to such silly ideas as, "You must believe in the great juju in the sky" or

"You must kneel down and face east and pray five times a day." He worries that ideas like these then get passed down through generations, without anything impeding their progress. Even worse is the fact that those viruses that are really good at infiltrating minds are precisely those with the most awful and dangerous messages. "So, if the virus says, 'If you don't believe in this you will go to hell when you die,' that's a pretty potent threat, especially to a child."

To say the least, this is all pretty emotive with the talk of viruses, something we immediately think about negatively. Perhaps Dawkins is right—Lucretius probably has another poem coming on—but one would like a little more reason for thinking religion false. Oxford-based Justin Barrett offers a no less naturalistic argument than Dawkins, claiming that religion comes from the overactivity of what he calls "agency detection devices" (ADDs). "Our ADD suffers from some hyperactivity, making it prone to find agents around us, including supernatural ones, given fairly modest evidence of their presence. This tendency encourages the generation and spread of god concepts and other religious concepts."[21] Interestingly and pertinently, however, Barrett is a committed Christian thinking that this could all simply be God's way of getting religion naturalistically. "Suppose science produces a convincing account for why I think my wife loves me—should I then stop believing that she does?"

To disprove religion one needs to turn to reasons drawn from the realm of philosophy and theology, and perhaps anthropology, rather than from evolutionary biology. Most obviously there is the problem that there are so many religions making contradictory claims. Why should one believe the Christians rather than the Muslims or the Jews or the Buddhists or the pagans? Why should one believe the Catholics rather than the Mormons? John Calvin, and following him Alvin Plantinga,[22] says that his religion carries the mark of its own authenticity, but we have heard that before—from just about every other religion. Then, compounding negative issues, with respect to Christianity there are already-raised problems about melding its Greek and its

Jewish roots. Is God a necessary being, outside time and space, eternal? Or is God a person, like the father in the story of the prodigal son, who welcomes his long-lost son but who also has understanding and sympathy for the boy who stays home? The two conceptions don't fit well together, and sometimes lead to horrendous conclusions, as when Anselm tells us that God does not feel some of the most basic of human emotions: "For when thou beholdest us in our wretchedness, we experience the effect of compassion, but thou dost not experience the feeling."[23] Or when Aquinas says: "To sorrow, therefore, over the misery of others does not belong to God."[24] Many of us just don't want a God like that. Or indeed, a God who allows so much evil into the world. The Christian worships an all-powerful, all-loving God. What price love now? Of course, Christians have their answers. For the poet John Keats, for instance, our world is the "vale of soul making," where suffering and hardship ennoble us. Others find this and related responses inadequate. Does one even want to believe in a God who let Anne Frank die in Bergen-Belsen? If someone starts trotting out the old chestnut that God gave us the great gift of free will and this made moral evil possible, one can only stand in horror at a deity who thinks the free will of Heinrich Himmler outweighs the pain and suffering of Anne Frank, or of Sophie Scholl whose life ended on the guillotine, because she belonged to the White Rose group opposing Hitler.

Other religions may not have all of the problems of Christianity—a religion like Buddhism without a Creator God is already one step ahead in simply not needing an explanation of evil—but they are hardly without difficulties of their own. Philosophers have pointed out that it is difficult to know quite how one maintains continuity for the individual if in the middle of existence there is a gap—between death and the Day of Judgment. Who is to say that the first Michael Ruse, professor, is the same chap as the second Michael Ruse, trying to persuade Saint Peter to open the gate? It has been suggested that perhaps consciousness is the software to the hardware of our physical bodies and that

God, as it were, keeps us on file?[25] But what then is to stop him making two, three, or even a hundred copies of Michael Ruse? Dizzying thought. And if it is difficult to think of making a repeat human Michael Ruse, imagine the difficulties if Michael Ruse is now a codfish. Perhaps it is psychically satisfying to think that Adolf Hitler is now a dung beetle in a galaxy far, far away, as one might say, but does it really make much sense? Nor for that matter is there much more sense in the pagan practice of "drawing down the moon," where the witch goes into a trance and has the Moon Goddess speak through her. I have considerable sympathy for the pagans—they are gentle folk who take the environment very seriously—but what they believe has no more connection to reality than reading golden plates through one's hat in Upstate New York or riding off in the middle of the night on a magic horse to have a few words with God about how often we should pray every day.

Does Religion Have a Purpose?

The conclusion thus far is that if religion is false, it is not obvious that Darwinism—certainly not Darwinism alone—is able to show this. I am talking now of a fairly sophisticated religion, one that has gone beyond the need to insist on a literal worldwide flood and such things. I don't dismiss the importance of Darwinism (and evolution more generally). I suspect that problems like the historical authenticity of Adam and Eve raise more difficulties than most Christians realize, but there are very traditional answers to such problems. Orthodox Christianity has never bought into the Augustinian take on original sin, involving Christ's substitutionary atonement on the cross. Rather, it sees humans developmentally, in a state of becoming, and it is Christ's incarnation and sharing of death with us in solidarity that counts. A historical Adam and Eve are not demanded.

Even if religion be false, we still have purpose within the system. Taken literally, the characters in *David Copperfield*—Mr.

Micawber, Uriah Heep, Dora Spenlow—do not exist, but one still has purpose in the novel. David ran away to Dover to find his aunt. Aunt Betsy concealed the extent of her losses to test David. Mr. Peggoty set out to find his fallen niece, Little Emily. But if religion be false, there is a new range of purpose questions. Why did it start and why does it persist? Does it have a real purpose? Both Dawkins and Barrett in their ways suggest that religion started as a by-product of useful adaptations. This is a line of thought that goes back to before the coming of evolutionary thinking. In his *Natural History of Religion*, Hume wrote, "We find human faces in the moon, armies in the clouds; and by a natural propensity, if not corrected by experience and reflection, ascribe malice or good-will to everything, that hurts or pleases us."[26] In other words, religion begins in mistaken identification of the inanimate with the living—indeed, a point we saw made almost two millennia earlier by Lucretius. Darwin, who as a young man had read Hume's essay, argued something similar. By the time of the *Descent* in 1871, Darwin had slid into a comfortable agnosticism. He dealt with religion briskly, arguing that it was all a matter of chance and confusion, thinking that the "tendency in savages to imagine that natural objects and agencies are animated by spiritual or living essences" was illustrated by the mistaken actions of his dog (a beast, Darwin tells us, who is "a full-grown and very sensible animal"). Snoozing on the lawn, the dog was upset by a parasol moving in the wind. Going on the attack "every time that the parasol slightly moved, the dog growled fiercely and barked. He must, I think, have reasoned to himself in a rapid and unconscious manner, that movement without any apparent cause indicated the presence of some strange living agent, and that no stranger had a right to be on his territory."[27]

In line with this approach, recently anthropologist Scott Atran has proposed a similar kind of by-product explanation of religion. It is all to do with our mechanisms for detecting danger and showing fear. "Natural selection designs the agency-detection system to deal rapidly and economically with stimulus situations involving people and animals as wired to respond to fragmentary information under conditions of uncertainty, inciting perception of figures in the clouds, voices in the wind, lurking movements in the leaves, and emotions among interacting dots on a computer screen."[28] This kind of adaptation can all too easily go astray. "This hair-triggering of the agency-detection mechanism readily lends itself to supernatural interpretation of uncertain or anxiety-provoking events."[29]

Why does religion persist? Here most people turn to a functional explanation—one invoking purpose—more or less of the kind proposed by the great sociological pioneer Emile Durkheim. With religion, we have a culture binding people and helping people and giving hope to all. Durkheim wrote, "A religion is a unified system of beliefs and practices relative to sacred things, i.e., things set apart and forbidden—beliefs and practices which unite in one single moral community called a Church, all those who adhere to them."[30] Giving this an evolutionary spin, Edward O. Wilson—no believer but much more sympathetic to religion than many—thinks religion is adaptive because of its power to confer group membership. "In the midst of the chaotic and potentially disorienting experiences each person undergoes daily, religion classifies him, provides him with unquestioned membership in a group claiming great powers, and by this means gives him a driving purpose in life compatible with his self interest."[31] Wilson does admit that there may be something to cultural causes, but essentially he thinks that it all comes back to biology. "Because religious practices are remote from the genes during the development of individual human beings, they may vary widely during cultural development. It is even possible for groups, such as the Shakers, to adopt conventions that reduce genetic fitness for as long as one or a few generations. But over many generations, the underlying genes will pay for their permissiveness by declining in the population as a whole."[32] Culture can play variations on the themes, but ultimately these themes are biological.

Is any of this well taken? One feels that there must be something to this way of thinking. Religions are such a prominent feature of human cultures, it would be very odd if they had no purpose at all, and conferring some kind of group solidarity seems as plausible as anything. It is not essential. Britain did not stand alone against the Third Reich in 1940 because of the Thirty-Nine Articles of the Church of England. For all that, religion can be important and a positive force. In line with what has been discussed earlier, historians have long made the case that Protestant Christianity was tremendously significant in the eighteenth and nineteenth centuries in defining and giving a sense of self-worth to Britons against the powerful forces on the continent.[33] Perhaps, even in 1940, the Church had its role in national pride and fortitude in making the V-sign to the Jerries, as the Germans were known. Many, like Darwin himself—Durkheim spoke of a "moral community"—thought religion essential to articulating and bolstering morality. One may perhaps have less confidence in this. Scandinavian countries, where religion has notoriously withered on the vine, score significantly higher on levels of well-being (including moral well-being) than countries with high levels of religiosity.[34] To take just murder rates: El Salvador (homicide rate of 71 per 100,000 inhabitants), Colombia (33 per 100,000 inhabitants), Brazil (26 per 100,000), and Mexico (18 per 100,000); Sweden, Japan, Norway, and the Netherlands (all with homicide rates that are less than 1 per 100,000). From a moral viewpoint, the American North is significantly more caring than the American South, and yet it is in the South where excessive evangelicalism thrives. No big surprise, for too often evangelicals spend time promoting the hate-filled prohibitions of the Old Testament rather than the love-filled prescriptions of the New.

One could go on searching for functional attributes of religion. Surely, with refined sentience giving the knowledge of personal death, the promises of religion have been important. The important thing is even if religion is false, there are many reasons to think that it generates enough purpose for its survival. This is not to say that, as happens with adaptations sometimes—one thinks of the peacock's tail feathers—it might not overstep the mark. Sometimes it is positively counterproductive. In the name of Jesus, priests and pastors on both sides of the trenches in the Great War urged young men on to their deaths in Flanders. Few equaled the truly dreadful Arthur Winnington-Ingram, bishop of London, who (in a 1915 sermon) urged his congregation "to kill Germans: to kill them, not for the sake of killing, but to save the world; to kill the good as well as the bad, to kill the young men as well as the old, to kill those who have shown kindness to our wounded as well as those fiends who crucified the Canadian sergeant."[35] But they were all cut from the same moral and theological cloth. One hardly has to be an enthusiastic eugenicist to think that killing off the best and brightest is probably not the best way to improve the human gene pool.

More recently, the dreadful instances of sexual abuse by the Catholic clergy suggest that group cohesion is not prominent in future prospects for that religion's survival. It is hard to think of purposes and values at a point like this. Although the human power of self-deception never ceases to amaze.

> What? I? "Ruined their lives"?
> Wait a minute, let's get this straight—
> my passion *gave* them a life, gave them
> something rich and ripe in their green youth,
> something to measure all intimate flesh against,
> forever. After that,
> they ruined their own lives, maybe.
> But with me they were full of a love
> firmer than anything their meager years
> had ever tasted.[36]

CHAPTER TWELVE

The End

Darwinism Again: Knowledge

I do not want to end all discussion of religion. Anything but. Here, in line with the sentiment expressed in my preface, I am more interested in stressing the positive than pushing the negative. Rather than spending time about why religion is wrong, I want to open the possibility of a life without religion, without God. Can one then have purpose, or is life all an empty charade? In the words of Ivan Karamazov: "Without God and the future life? It means everything is permitted now, one can do anything." And if everything is permitted, then nothing has any special value, and as we have seen, value is at the center of purpose. Life is without purpose.

Many people think of Darwinism as an alternative religion. Julian Huxley actually wrote a book called *Religion without Revelation*. Edward O. Wilson is of the same mind-set. Anyone who knows their scriptures has to be forcibly reminded of the Old Testament prophets on reading Richard Dawkins's *The God Delusion*. Others, like myself, prefer not to go down this path. Having given up our childhood faith, we do not want to take it up again even in a secular fashion. We shudder at celebrating Darwin's birthday and calling it "Darwin Day." The next thing is they will be putting him in a manger. That does not mean that evolution, Darwinism, cannot help with finding alternatives, and indeed, if you think (in the words of Thomas Henry Huxley) that we are modified monkeys rather than modified dirt, it is nigh compelling to turn to Darwinism for help. We are giving up one story of origins, so it is natural to turn to the alternative story of origins. In this sense, evolutionary thinking is privileged over (let us say) organic chemistry. And to the naysayers like Thomas Nagel, I can but quote John Stuart Mill: "And if the fool, or the pig, are of a different opinion, it is because they only know their own side of the question. The other party to the comparison knows both sides."[1]

So how do we set about the task? We use our minds to think, to reason. What are we thinking or reasoning about? Let us agree (with qualifications to come) that ultimately we are thinking and reasoning about things that will help us successfully to survive and reproduce. But what as animals—particularly what as humans—do we need or do to survive and reproduce? Kant is helpful: "Two things fill the mind with ever-increasing wonder and awe, the more often and the more intensely the mind of thought is drawn to them: the starry heavens above me and the moral law within me."[2] In other words, knowing about the world around us and having a moral sense that guides us in our relationships with others, especially other human beings. Let us explore these two points.

If you stand in the Judeo-Christian tradition, you know—or at least you can know—truly about the physical world (including the living world) in which you live. You are made in the image of God, and while you may be tainted by original sin, there are going to be methods to get at the truth. Descartes, remember, suggested that we can discern clear and distinct ideas and that they tell us truthfully about what is guaranteed by God. If you are a Darwinian evolutionist, then things get a little more complicated. A fairly standard view (to which I subscribe) of Darwinian evolution at work on problems of knowledge—what has been given the rather ugly name of "evolutionary epistemology"—sees knowledge structured by innate dispositions about reasoning and mathematics and so forth, what Kant locates at work in the synthetic a priori, but with these dispositions having been put in place by natural selection for their utility.[3] They are not, as Kant thought, necessary conditions for all and any rationality. The dispositions are then filled in, as it were, by experience and culture. Darwin, for instance, in the *Origin*, made use of a consilience, which is a method of argumentation that because of its utility was put in place by selection, but the details were filled in by experience (as on the Galapagos) and culture (as in using metaphors like a division of labor).

Is this enough? The sometime English prime minister Arthur J. Balfour (1848–1930) argued strenuously that natural selection is a poor reed on which to put one's faith for truly discerning the nature of reality:

> We are to suppose that powers which were evolved in primitive man and his animal progenitors in order that they might kill with success and marry in security, are on that account fitted to explore the secrets of the universe. We are to suppose that the fundamental beliefs on which these powers of reasoning are to be exercised reflect with sufficient precision aspects of reality, though they were produced in the main by physiological processes which date from a stage of development when the only curiosities which had to be satisfied were those of fear and those of hunger. The instruments of research constructed solely for uses like these cannot be expected to supply us with a metaphysic or a theology, is to say far too little. They cannot be expected to give us any general view even of the phenomenal world, or to do more than guide us in comparative safety from the satisfaction of one useful appetite to the satisfaction of another.[4]

Actually, a version of this argument was raised by the atomist Democritus. He was an empiricist, wanting to explain only in terms of the sensed. But as an atomist, he realized that his senses must be deceiving him, for the world he sensed was solid and colored and so forth, not buzzing little balls. So since his empiricism was not reliable, how then could one infer anything? In the words of Galen (129–ca. 200), the Greek physician, Democritus has his senses say to his intellect, "Wretched mind, do you take your evidence from us and then try to overthrow us? Our overthrow is your downfall."[5] In recent years, Alvin Plantinga has followed a similar line of reasoning, arguing that we cannot rely on processes that evolved solely for the purposes of survival and reproduction. Natural selection could mislead us for our own biological good and we could be living in a state of total deception. Somewhat cutely referring to what he calls "Darwin's Doubt," because it was a worry expressed by Darwin himself ("With me the horrid doubt always arises whether the convictions of man's mind, which have been developed from the mind of the lower animals, are of any value or are at all trustworthy. Would anyone trust in the convictions of a monkey's mind, if there are any convictions in such a mind?"[6]), Plantinga inventively pretends to be present at a posh dinner in an Oxford College, where Richard Dawkins is arguing for atheism before the philosopher A. J. Ayer—a classic case of coals to Newcastle, one would have thought. Perhaps biologist and philosopher are living in a dreamworld. Their beliefs "might be like a sort of decoration that isn't involved in the causal chain leading to action. Their waking beliefs might be no more causally efficacious, with respect to their behaviour, than our dream beliefs are with respect to ours. This could go by way of pleiotropy: genes that code for traits important to survival also code for consciousness and belief; but the latter don't figure into the ethology of action. It *could* be that one of these creatures believes that he is at that elegant, bibulous Oxford dinner, when in fact he is slogging his way through some primeval swamp, desperately fighting off hungry crocodiles."[7] Natural selection could be making a sham of everything we believe about the world of reality.

Obviously, a lot of what these critics are saying is true. We could be living in a fool's fantasyland. And if what people like Hume tell us is true, we do project a lot into the real world and think then that we have read it off. Causal necessity for a start. Perhaps religion for a second. But notice that selection does not leave us totally helpless. We can often have a pretty good idea of when nature is deceiving us. The burned child fears the fire. Beings that see a fire and associate with it burning and pain and the like are adaptively ahead of those who say, "Fire burns us? Just a theory, not a fact." I made mention earlier of W.V.O. Quine, who knew the score: "If people's innate spacing of qualities is a gene-linked trait, then the spacing that has made for the most successful inductions will have tended to predominate through natural selection. Creatures inveterately wrong in their inductions have a pathetic but praise-worthy tendency to die before reproducing their kind."[8] The fact is that if you are fighting crocodiles, then what you need are skill and cunning and energy. Boozing it up with Freddie Ayer is not the key to success of that sort.

Balfour is right. It is remarkable that our adaptations do so much. But if they started off by telling us about the world, then I am not sure why they should not go on telling us about the world. No one is saying, for instance, that selection gave us a gene for understanding quantum mechanics straight off. However, if it did give us genes for straightforward observation and reasoning, that is basically all one can or need ask for. The critics do point to the fact that ultimately for the Darwinian, it is a matter of getting it all to hang together. If the Humean analysis of causation fits, then plug it in. I worry in major part about Christianity because I cannot reconcile Athens and Jerusalem. Plantinga objects that this means we still in principle could be overall mistaken, just like the man in the factory who (unknown to himself is wearing red-colored glasses) sees everything as red and thinks this is so even if it is not. He has no means to judge outside the system. This is probably true and points ultimately to the fact that truth for the Darwinian is coherence rather than correspondence—it can be correspondence within the system but not overall. This, one would say, is the human existential position. Is the Christian any better off? Descartes and Plantinga (following Calvin) think that God guarantees truth. Perhaps, alas, Descartes's evil demon who corrodes everything, once let out of the bottle, can never be recaptured. Can one ever be absolutely certain that one is not being deceived, especially given that others, equally certain, believe other things?

As we prepare to move on now to morality, note that in major respects our purpose-driven lives (if we may borrow a phrase from the title of a book by an evangelical who would agree with absolutely nothing in this book) come from the fact that we can tell something about reality. Because I am hungry and I can see animals out on the plain, it makes sense to plan and devise ways in which I can catch them and eat them. Because Emily had seen lawyers at work and down the road visited public defenders' offices, it made sense for her to strive to join such an office herself. Note, however, that there is nothing to say that everything we do purposefully has to be tied directly to survival and reproduction. This is the thing about culture, the product of our minds and our reasoning and our efforts based on these: On the one hand, it is an incredibly powerful new way of transmitting information for our own ends. There are reasons why a naked ape from Africa lives all over the world in ever-increasing numbers. Someone has a breakthrough in agriculture and you don't have to wait for the right genes to keep appearing and for selection to distribute them. The ideas can be passed on quickly from grown-up to grown-up. On the other hand, there is somewhat of a decoupling from survival and reproduction. One can well imagine that a fondness for games and physical play has biological adaptive virtues, but it is hard to imagine that American college football has any such virtues. The very opposite, in fact, what with the damage done to the bodies of young men and the drinking that goes on among spectators on football weekends in the fall. And I don't even want to get into the moral corruption of a supposed institution of higher education that pays its football coach one hundred times what it pays an assistant professor. At a more refined level, it is a commonplace that the most esoteric flights of pure mathematics have a way of finding practical applications, but there is certainly not a priori reason why finding the Euler identity should do anything to improve anyone's survival and reproductive chances. Something in culture must be adaptive, or we wouldn't be here, certainly not in such numbers. There is no reason for everything to be adaptive, especially if it is not positively counteradaptive. The Shakers are now known for their furniture, not their megachurches.

Darwinism Again: Morality

What about morality? Go at it backward. It is absolutely and completely teleological. It is a major reason why there is purpose in our lives. I am sitting around the living room on a Saturday afternoon watching college football. I ought to be out on the green playing soccer with my sons and my daughter. I am staying up for three nights and drinking nonstop so I will be rejected at my medical, but I ought to be signing up to fight Hitler. I live in Florida and fly to Europe at least six times a year. I am adding considerably to factors causing global warming. I should quit my job, move to North Dakota, join a commune, and live in a yurt, eat only raw vegetables, go everywhere on foot or on a bicycle, and only have sex with my handkerchief lest I add to the population explosion. We are always thinking in terms of ends, of purposes, and of what we should be doing now and what we should not be doing now. As always, it is a matter of value. It is better to play with my kids than to vegetate on a couch in front of the television. It is better to fight Hitler than cowardly to avoid the responsibility. It is better to munch carrots in the wilderness by the Canadian border than to sit in a café on the left bank of the Seine with a glass of red wine and a smidgen of brie, arguing about Michel Foucault with Parisian pseuds like myself.

We can skip over solutions like that of Moore that lie in the Platonic tradition and equally over solutions that lie in the Aristotelian tradition, although as noted earlier, Aristotle-inspired ethics, so-called virtue ethics, finds many supporters today. What of the Darwinian case? We must tread carefully here, for since the *Origin*, far better known has been the position on ethics of Darwin's fellow British evolutionist Herbert Spencer. And as one starts to dig into Spencer's thinking, one starts to think that perhaps the British philosophers had a very good point. Stay away from this kind of stuff! At the normative or descriptive level—what should I do—the early Spencer can sound positively brutal about letting widows and children go to the wall.[9] At the level of justification, metaethics—Why should I do what I ought to do?—one's sympathies are with G. E. Moore. A truck is driven through the "is/ought" distinction. "Ethics has for its subject-matter, that form which universal conduct assumes during the last stages of its evolution."[10] Then: "And there has followed the corollary that conduct gains ethical sanction in proportion as the activities, becoming less and less militant and more and more industrial, are such as do not necessitate mutual injury or hindrance, but consist with, and are furthered by, co-operation and mutual aid."

If you think this is bad enough, let me ruin your day entirely by telling you from Spencer to the present, there have been those (usually biologists rather than philosophers) who have happily and proudly followed in the tradition. Above all, it is to be found today in the writings of Edward O. Wilson, for which enthusiasm he has long been the object of condescending scorn from members of the philosophical community, including, I confess, myself.

> While many substantial gains have been made in our understanding of the nature of moral thought and action, insufficient use has been made of knowledge of the brain and its evolution. Beliefs in extrasomatic moral truths and in an absolute is/ought barrier are wrong. Moral premises relate only

to our physical nature and are the result of an idiosyncratic genetic history—a history which is nevertheless powerful and general enough within the human species to form working codes. The time has come to turn moral philosophy into an applied science because, as the geneticist Hermann J. Muller urged in 1959, 100 years without Darwin are enough.[11]

For Wilson, humans have evolved in symbiotic relationship with the rest of the living world, and if we destroy that world, we destroy ourselves. This is why, in the name of evolution, he has become an ardent spokesperson for "biophilia," arguing that unless we save such entities as the Brazilian rain forests, we are doomed.[12]

Now I am certainly not about to launch a full-blown defense of Spencer, although it is worth noting that at the normative level, his kind of thinking does not necessarily commit one to a laissez-faire morality that would do credit to Margaret Thatcher—a name, incidentally, not chosen at random, for she came from the same British Midlands, lower-middle-class, nonconformist background as did Spencer, and she, like Spencer, was less interested in having widows and orphans starve than in breaking down the powers of the traditionally ensconced rich and powerful. Wilson shows us that there are more acceptable normative claims that one can embrace, and Spencer himself was a major voice for free trade between nations and the hope thereby of ongoing peace. At the metaethical level, there is no question that Spencer and his followers do smash through the is/ought distinction. The question though is what precisely this means and entails. What if you deny the validity of the is/ought distinction and argue in some sense that physical things, including organisms, have some kind of absolute value in themselves? If this is so, then seeing ever greater value emerge is almost to be expected.[13] We have already seen people who think this way—Plato and Aristotle with their organic analogies, for a start. Remember that Spencer himself, although he was always loath to admit intellectual debts, owed much (via the writings of Coleridge) to Friedrich Schelling, the Romantics' Romantic. He in turn owed much to his predecessors—Spinoza, Aristotle, and even more to the side of Plato, on which we are now focusing. Mention has already been made of that juvenile, sixty-page essay on the *Timaeus*.

I am not unaware of the paradox of saying that there are Platonic elements in Spencer's thinking, having earlier said that his greatest critic, Moore, is also a Platonist. As Whitehead said, all philosophy is footnotes to Plato. There are different sides to Plato and that is what is at issue here. This does not now mean that I am endorsing Spencer; rather, pleading for a more sympathetic understanding. The most beautiful place in the world is the Stellenbosch wine-growing area in South Africa. If some mining company moved in, intending to tear off the tops of the mountains, I would be ahead even of the ecofeminists in crying "rape"—and if that is not a value cry, one made for the sake of the mountain and not for me, I don't know what is. Returning to Spencer, more importantly for us here, I am saying he is not really in the Darwinian tradition but more in that of the Greeks. Although I suspect that if your Darwinism pushes you toward monism, especially toward some form of panpsychism, seeing mind as all-pervasive, then a spirited case might be for saying that a Darwinian could and should go a long way down this path. As goes mind, so follows value—although the counter might be that while this is true for full-blooded panpsychism, it does not necessarily follow for a weaker form. Transferring information instantaneously across huge distances may make you inclined to think that mind is all-pervasive. Whether this transfer is something of value is another matter.

Leaving this, what about (what we might call) a more direct Darwinian approach to morality? One that preserves its purpose-laden nature? True confession time. I was the coauthor of the just-quoted, neo-Spencerian passage by Wilson! However, where he read the passage as saying that we can push through the is/

ought barrier and use nature to justify morality, somewhat disingenuously, I meant that we could do an end run around the barrier and use evolution to explain away the metaethical justification of normative ethics. Endorsing what has become known as the "debunking" argument, I argue that once you have given a Darwinian explanation of moral beliefs, you see that there is no foundation. Morality is a set of subjective beliefs, not a reflection of objective, human-independent reality.[14] To quote our heavy-booted coauthors again: "Ethics is an illusion put in place by natural selection to make us good cooperators."[15]

I will skip quickly over the science that shows that morality is something that emerges from the workings of natural selection. Although there is still much controversy about how exactly natural selection does this, it seems generally agreed that cooperation—altruism—is something of value to the group and via this to the individual.[16] The words of Darwin still stand today: "It must not be forgotten that although a high standard of morality gives but a slight or no advantage to each individual man and his children over the other men of the same tribe, yet that an advancement in the standard of morality and an increase in the number of well-endowed men will certainly give an immense advantage to one tribe over another."[17] He continues: "There can be no doubt that a tribe including many members who, from possessing in a high degree the spirit of patriotism, fidelity, obedience, courage, and sympathy, were always ready to give aid to each other and to sacrifice themselves for the common good, would be victorious over most other tribes; and this would be natural selection."[18] Hence: "At all times throughout the world tribes have supplanted other tribes; and as morality is one element in their success, the standard of morality and the number of well-endowed men will thus everywhere tend to rise and increase."[19]

Even though this may explain why we think morally, why does evolution show that there really are no foundations? If evolutionarily evolved adaptations can tell us truly about the physical world, why can they not do the same about the moral world? The reason is that, if a speeding train is bearing down on you, you had better get out of its way. It doesn't really matter how you get to know it. If insect-like chemical sensors or bat-like echolocation did the job better than sight and sound, we would surely have evolved in a different way. But to the same end. Morality is similar and yet more radical. Yes, there is the "same end," but whereas in epistemology it is about something, the train, really "out there," in ethics it is about human relations and getting on. There is not the physical "out thereness" of the train, and it is here that the nondirectionality of evolution really kicks in—something about which most philosophers are nigh deliberately obtuse. If you could reproduce more by being Attila the Hun incarnate, natural selection would push you that way. By and large, however, that is not a genuine option for most people, and so we have been shoved toward some form of cooperation. There is no Seal of Good Housekeeping on which way. This lays open the possibility that, as opposed to what we do have, one could have a completely different yet functioning moral code—or a substitute for a moral code. If the aim of morality is getting along with each other, Kant allows that we might have no morality at all and just work through self-interest. "What concern of mine is it? Let each one be as happy as heaven wills, or as he can make himself; I won't take anything from him or even envy him; but I have no desire to contribute to his welfare or help him in time of need."[20]

Kant does agree that in real life this wouldn't go too far because it takes out the human need of sympathy and feeling. But—even if we agree with what does seem implicit in Kant that we must obey some formal rules of reciprocation—we can imagine fairly humanlike creatures with emotions and a different moral system. Suppose that, rather like John Foster Dulles (President Eisenhower's secretary of state), in the 1950s dealing with the Russians, instead of thinking that one should love one's enemies, one thought one should hate one's enemies—moral obligation. However, one knew they felt the same about you and so you got on—as did Dulles and the Russians. So now you have two functioning moral codes. It just so happens you have developed one rather than the other. You could have developed the other. Which is the true one? Who can say? And before you say that at least there was one true code, notice that its truth seems inessential to your belief system, and that is surely antithetical to what we understand by objective moral standards. It is certainly antithetical to what Moore understood.

By pointing out the consequences, if we had evolved in a different way, Darwin was even more radical: "If, for instance, to take an extreme case, men were reared under precisely the same conditions as hive-bees, there can hardly be a doubt that our unmarried females would, like the worker-bees, think it a sacred duty to kill their brothers, and mothers would strive to kill their fertile daughters; and no one would think of interfering." Continuing: "The one course ought to have been followed, and the other ought not; the one would have been right and the other wrong."[21] Supposing the nonworking brothers to have devoted their time to intellectual study, on a regular basis one would have had female drudges killing off male Aristotelian philosophers, all in the name of morality. Hmm.

Here is not the place for detailed defense of the Darwinian position just sketched. Let me simply make two points. The first is that, if the argument is well taken, it does not mean that substantive morality now vanishes or collapses. It is very much a position in the tradition of Hume and the other eighteenth-century empiricists on down to the logical positivists of the twentieth century and the "emotivism" that emerged from this. The attack is on foundations—be these God's will, or Platonic forms, or Moore's nonnatural properties, or, indeed, the natural properties of Spencer and Wilson. Although it is less distant from Kant, whose making ethical norms part of the synthetic a priori meant that they come from us rather than found "out there," as with epistemology there is a subjectivity that is denied by the Kantian necessary conditions for any rational being to think and act. As Darwin pointed out, there is a kind of evolutionary relativity—different moral codes for different kinds of being—yet since *Homo sapiens* is all one species, for us there is not that much moral relativity, and such as there is probably more cultural than biological. Aristotle thought it morally acceptable to have slaves; we do not. The change is one of culture and not of genes.

All of this means that purpose talk is proper and meaningful. My wife and I gave Emily money every month so she could work pro bono at the public defender's office, with the aim of her eventually getting taken on as a paid employee. Because she was our daughter, we had a special moral obligation to her, to see that she had a good start in life and that she herself could grow up into a well-rounded person, able to make a proper contribution to society. We paid out then, with the purpose of achieving in the future something we thought morally important. The belief that we had such an obligation to our children is part of our moral code or system. These are our values. This deliberately chosen example does emphasize that an evolutionarily based—not justified—morality will have a distinctive form. It would see an obligation to all in need, but would argue that we have special obligations to some—our children and other relatives particularly. Although Peter Singer might dispute this,[22] Saint Paul would not. "I seek not yours but you: for the children ought not to lay up for the parents, but the parents for the children" (2 Cor. 12:14). Nor would Hume: "A man naturally loves his children better than his nephews, his nephews better than his cousins, his cousins better than strangers, where everything else is equal. Hence arise our common measures of duty, in preferring the one to the other. Our sense of duty always follows the common and natural course of our passions."[23]

The second point is that, with morality, we have a paradigmatic case of evolution deceiving us for our own good. "Reason is as cunning as it is powerful. Cunning may be said to lie in the intermediative action which, while it permits the objects to follow their own bent and act upon one another till they waste away, and does not itself directly interfere in the process, is nevertheless only working out its own aims."[24] I don't suppose I was alone who, being introduced to moral philosophy more than fifty years ago, found the then-popular ethical philosophy of emotivism dissatisfying to the point of immorality. It said that claims like "Rape is wrong" translate out as "I don't like rape. Boo hoo! Don't you like it either?" (Refinements like "prescriptivism" added things like "Don't rape.") This could not be so. "Rape is wrong" means rape is wrong—it is morally prohibited—even if the whole world thinks it is okay. It was wrong to be prejudiced against Jews even though 80 percent of Germans under the Third Reich thought it acceptable. What was missing in the analysis, as people like John Mackie pointed out, was the sense of absoluteness. The meaning of moral statements includes objectivity.[25] "Rape is wrong" means it is objectively wrong to rape. And it doesn't take much to see why evolution added this element to the pie. If we thought it was all feeling, then the temptation to cheat would be overwhelming and substantive ethics would break down almost immediately. Because we think morality is binding on us, we do not cheat—at least, if we do cheat, we know that it is wrong. Before we have finished, we will be picking up again on some of these issues, but as we move on, let us collect what we have. A Darwinian evolutionist can and does have moral purposes. Generally, these will be the same purposes as everyone else—"don't sexually abuse small children"—but although we don't have extreme relativity, they will be geared to our underlying biology. One could never think (all other things being equal) that strangers are more important than family. Ultimately, moral purposes are part of the human condition, not existing outside us. I would speak of this, in the terms of this book, as more in the Kantian tradition, except it is not really Kant's own position. You know what I mean, so let us leave it at that. There are values. There can be purposes.

Cultural Progress?

How then are we to tackle the Dostoevsky problem without God? If God does not exist, can life have any purpose, any meaning? Take first the historical dimension. I suspect I am not alone when it comes to thinking about the secular notion of purpose through history, the idea of social or cultural progress—something with a goal toward which history is directed. Clearly we can make a case for progress if we think of things brought about thanks to science and technology—for instance, the Internet and how in the lifetimes of most of us it has transformed the way we think and work. Medicine too. Think of how smallpox has been wiped out and how polio is on the brink of extinction. Yet balancing this are horrendous conflicts—two world wars in the twentieth century for a start—as well as other massive acts of cruelty: Stalin and the kulaks in the 1920s and 1930s and Hitler and the Jews in the 1930s and 1940s. Steven Pinker argues that, despite these and other acts of violence, the world nevertheless is becoming a friendlier place.[26] Perhaps, although I suspect many of us think a utilitarian head count is not adequate or appropriate, and that each and all of these horrors makes talk of progress not just otiose but somewhat obscene. Inexcusably naive too, faced as we are now with the spread of nuclear weapons to such unstable regimes as North Korea and the seemingly inevitable global warming. It would take a foolish and dangerous optimist to speak confidently of ongoing comprehensive cultural progress. Science and technology seem as much the problem as the cure.

This said, one can see progress in limited areas, and not just in science and technology. Think, for instance, of women's education, at least in the West, in the past hundred years. When I was born, nice girls headed to secretarial school. No longer. And so it surely makes sense to think in terms of purpose, at some kind of collective, historical level. About fifty years ago, the country in which I have lived most of my life, Canada, introduced government-sponsored, universal health care. This did not come about by chance, nor is it universally popular by chance. People set out to start it and others to continue it, because they thought and still think it a most worthwhile end. One can have purpose and one can achieve these ends. One does not have to be quite the disillusioned cynic that *Candide* becomes at the end of Voltaire's novel, that the best we can do is stay home and tend our garden. That said, grandiose Enlightenment schemes, modeled as they were on Christian promises of eternal bliss, seem far-fetched and far away. It is interesting and instructive how much of the global warming debate on both sides is carried on in apocalyptic terms. Arguably, one of the deadliest legacies of Christianity is to incline us to think of history purpose-driven to a desired end.

Personal Meaning

At the individual, personal level, what of purpose for the nonbeliever? When I lost my faith around the age of twenty, I was not at all sure that such purpose was possible. And in a sense, of course, I was right. I had given up the idea of purpose offered by Christianity, namely, striving in this world for rewards in the next. Of course, with reason you might respond that it is as well that I gave up this idea of purpose, because it certainly isn't that of Christianity. Jesus knew full well the joys as well as the sorrows of this life. Think of his friendship with the disciples and with Mary and Martha. This said, there is more than a flavor of this end-direction about Christianity and religion in general. The problem of evil is explained this way. Cancer in the child is made understandable by God's plans for the hereafter. Kant made that point about truth telling. Never telling lies can lead to horrendous problems; fortunately, God will make it all right in the long-term, and we must always keep this in mind. Returning to nonbelievers like me, nothing denies that one can have a lifelong purpose aimed at a goal before this life ends, for instance, making a billion dollars. Or, perhaps more elevated, for Zionists seeing the creation of the state of Israel. But I suspect that most of us, having given up the idea of a lifelong purpose aimed at the next world, are inclined to draw back somewhat and cut down on lifelong purposes generally. After all, if everything you do and think is fixed on your next decades, then are you not missing out on the decade you are in? For the nonbeliever, these are all you are going to get. My existence, the value to my life, was not going to be predicated on the hope of feeling satisfied as life draws to an end—although as it happens, I do feel satisfied—but rather on doing those things of value along the way that will lead to such satisfaction. This all starts to sound very Aristotelian, and I think it is, so long as one doesn't try to read too much overall meaning into things. In other words, as long as one is first and foremost a Darwinian! This said, as virtue ethicists stress, one pulls back from grandiose plans and one cultivates the things that are important to you as a human being. When I say "cultivate," there is obviously some real intention and thought here, but a lot of it is actually doing and trying to bring meaning and value—and purpose—into one's life.

My fellow philosophers have written intelligently and sensitively on these matters and help me to see things in perspective. I have always found inspirational Jean Paul Sartre's little essay *Existentialism Is a Humanism*, based on a lecture given in 1945. He tells us that existence precedes essence and that, in a world without God, we must do the creating ourselves. "There is no human nature, since there is no God to conceive of it. Man is not only that which he conceives himself to be, but that which he wills himself to be, and since he conceives of himself only after he exists, just as he wills himself to be after being thrown into existence, man is nothing other than what he makes of himself."[27] Continuing: "What we mean to say is that man first exists; that is that man primarily exists—that man is, before all else, something that projects itself into a future, and is conscious of doing so. Man is indeed a project that has a subjective existence, rather unlike that of a patch of moss, a spreading fungus or a cauliflower. Prior to that projection of the self, nothing exists, not even in divine intelligence, and man shall attain existence only when he is what he projects himself to be—not what he would like to be."[28]

I think there is a human nature—our knowledge and our morality—one that I have been sketching earlier in this chapter, one that was shaped by Darwinian factors. That is the beginning of freedom, not its end. Trying to cash out how one now moves forward, how one sets about creating oneself, turn to the insights of American ethicist Susan Wolf, who sees meaning in fulfillment—"one finds one's passion and goes for it"—and in going beyond self, "a life is meaningful insofar as it contributes to something larger than itself," with the proviso that this circles back to self: one has "an expectation about the subjective feelings and attitudes that contributing to something larger will engender."[29] She writes also that "our susceptibility to these sorts of reasons is connected to the possibility that we lead meaningful lives, understanding meaningfulness as an attribute lives can have that is not reducible to or subsumable under either happiness, as it is ordinarily understood, or morality."[30] I am not sure about this. If you understand "happiness" in the extended sense of John Stuart Mill—better Socrates dissatisfied than a fool satisfied—then I would argue that meaning and happiness do go together. One of the commentators on Wolf raises the case of Claus von Stauffenberg who led the plot against Hitler. He was hardly cheery when the plot failed, when he was discovered and was about to be put to death. But his life was surely meaningful, and even at that time—especially at that time—he had a sense of self-worth, which is the mark of the truly happy person. With respect to morality, as a Darwinian I want to get away from the Christian notion that you are either in or out on the issue. Prison visiting is of moral worth; composing operas is not. Apart from endorsing the Kantian notion of duties to oneself—Mozart had the duty to use his phenomenal gifts—I see value more on a spectrum, which of course is what you might expect from an evolutionist. I am not sure you could have a totally meaningful life if one were totally selfish—Richard Wagner around your wife, for instance—but value slips easily from one end of the spectrum to the other. Mozart and Wagner have brought great happiness to their fellow human beings, and that is surely a morally good thing to have done—quite apart from the worth for themselves of composing. Wolf insists on morality in some sense being objective, but "objective" is a very mild term for her; she stresses explicitly that she is not demanding nonnatural, Platonic-like qualities of the kind supposed by G. E. Moore, and if we mean by "objective" going beyond the purely relative—if it feels okay, then it is okay—as we have seen, the Darwinian insists on this. Although, it is comparative value that is at stake, when dealing with organisms, the human realm alters this. There might be some discussion about the absolute value of using an automobile rather than a bicycle—this is part of the discussion about whether culture shows real progress—for all that there are deniers, it is surely legitimate to say (as I have said) that vaccination against smallpox or polio is an absolute for the Darwinian as much as for anyone.

At the risk of being even more egocentric than usual, using the excuse that this is all personal and there are no outside supports, let me talk of three things that have given purpose to my life—made my life meaningful in a sense that I think would be appreciated by Wolf and others. First, family. After a not-very-happy first marriage, and several years as a single parent, I met my present wife, Lizzie, who, fortunately, for all that she shares her birthday with Beethoven, is not named "Ludwig." We have had more than thirty years together raising children, not always finding it easy but fortunately having enough shared sense of humor to get through to the next day. Now with the kids launched—more or less, some days rather less than more—we find we really like each other and go traveling and those sorts of things. She regrets the books that keep piling up. I am seriously thinking of joining Amazon Anonymous—"Hi, I'm Michael, and I haven't bought a book in five days." I regret the dogs she keeps bringing home—"Hi, I'm Lizzie, and I haven't been to the Bainbridge Animal Shelter in five days." But we compromise. I buy books instead from AbeBooks.com. Lizzie stays in-state for her dogs. Nutmeg is a whippet from a Florida breeder. Our shared understanding and mutual tolerance, and the results that follow, are certainly things of value—things that give purpose to life. As Plato says in *The Republic*, they are the best kind of good—a good here and now and a good for the future. As are the children. The fun of the children at the time—am I the only person in the world who really loved having teenagers around the house?—and the joy that, like Emily, they have found ways to meaningful lives.

To family, in a very Greek way I would add friends. I am a compulsive worker, and as a break, I love to cook. What is good about this is that first, it is intense and you have to focus on what you are doing and not on other things. For a time, my mind is not racing on about the latest philosophical puzzle. It does not preclude listening to the radio. No account of my life would be complete without a tribute to the Metropolitan Opera and its Saturday afternoon broadcasts—now, and this starts to make even me think there might be a Good God, supplemented by cinema showings of matinees. Send out for pizza on those days! The second thing about cooking, or rather its results, is that it is social—blurring the distinction between value for oneself and value for others. Food is for sharing and conversation and more. At least, now that the second batch of children is in their twenties, we no longer have to buy loaves by the dozen and potatoes by the hundredweight. Although, be warned. Our youngest, Edward, is in Britain doing graduate work in philosophy. "Friday the Thirteenth. Just when You Thought the Worst Was Over! Fifty More Years of Philosophical Ruses."

The second source of great satisfaction and purpose is being of service to others. That Quaker childhood struck deep! Sartre also: "When we say that man is responsible for himself, we do not mean that he is responsible only for his own individuality, but that he is responsible for all men."[31] I am absolutely not a do-gooder. The thought of going to Central America to build houses for the poor terrifies and appalls me. And I am not into late-night soup kitchens. But I have been a teacher—a college prof—now for fifty years and I find it deeply satisfying. I cannot say that I have always done brilliantly—my teaching ratings are pretty good, but I am not sure at all that I trust those—but I have striven to speak to every student, and (Quakerism again!) to see, in an entirely secular sense, that of God in every person, the "Inner Light." I should say that being a philosopher has been important here because philosophy is the highest calling. Plato was absolutely right about this. To be able to share this with young people has been a joy and a privilege. My undergraduate teaching has been very heavily geared toward first-year students. I had a hugely difficult time making the transition from the close atmosphere of a Quaker boarding school—my American friends do a double take when I tell them that I am the product of a Christian high school—to the rather alienating experience of university. I want to help young people know that they are not alone, and that although life can be challenging, it can also be exciting and rewarding. For better or for worse, I am sure my writing style is a function of all of this—never presuming, always trying to keep the audience's attention.

At the graduate level, most of my interaction has been in the second half of my career. In part, this was because in the early years I was establishing myself and wasn't really ready. I set myself a huge agenda as I moved from philosophy to biology and then on to the history of science. The social psychologist Donald Campbell once said that to be interdisciplinary, you have to be willing to be inadequate in many fields at the same time, and I know what he meant. As I steered into waters unknown, understandably, few students wanted to follow me. Apart from anything else, it was one thing for me, a tenured full professor to do this, another for a soon-to-be job applicant. I, too, worried, as I still do, about their job prospects. It isn't good enough to mentor and cherish someone rather vulnerable for five or more years and then turn around and say that you are not an employment agency. But I did want to contribute, and for this reason I started the journal *Biology and Philosophy*. Also, I have done a lot of book editing, in two series for Cambridge University Press. In that way, I was able to help others, particularly those at a junior level. It wasn't just a matter of doing good in a Mother Teresa fashion. As I think Susan Wolf would appreciate and I hope approve, it has been great fun, not the least because in the journal I ran a column called "Booknotes," where I had license to say what I wanted about a lot of self-regarding people. More recently, I have taken on graduate students, and I feel great pride in their development and successful job hunting, if not in academia, then in rewarding work elsewhere. As the students grow, particularly the graduate students, they turn from being one's children into being one's friends, and relationships forged are ongoing beyond college days—meaning in the sense of being fulfilling and meaning in the sense of being larger than oneself. I never, ever thought I would become a notary public and perform weddings, but I have and I did. I should say that all of the ethnic grandparents were greatly relieved when I used the service from the Book of Common Prayer. That I omitted all references to the deity mattered far less than the avoidance of a flower-children event, where the lines are made up—even worse, where people read from *The Prophet*—and we are all expected to sing Bob Dylan songs and embrace each other and the breakfast is gluten free.

Third, and this moves right on from the last point, the life of the mind, of the intellect. I always knew I was going to be a writer and I have been. The teacher in my infants' class read "Rikki-Tikki-Tavi," and I was hooked. I cannot remember when I could not read, and from the age of five or six I had my nose in a book, always. The dreaded observation that the weather is nice—fortunately we lived in England—was always a wrench as I was pulled away from *The Children of the New Forest* or *The Secret Garden* or *The Family from One End Street* or (although my sister's, one of my all-time favorites) *Ballet Shoes*. Thank you, Andrew Carnegie! As I have grown up, my tastes have changed, although the love of reading stays. Happiness is an old favorite by Charles Dickens or Anthony Trollope. I confess a weakness for the "shockers" of John Buchan, especially the tales of Dickson McCunn, the Glasgow grocer. Apart from his phenomenal storytelling powers—Mrs. Gaskell and Neville Shute also come to mind—I should say that what makes these tales particularly gratifying is that, under the cover of a fast-moving thriller, Buchan works to expound and understand the Calvinist theology he imbibed as a child. It is for much the same reason that I enjoy and admire the trilogy—*Gilead*, *Home*, and *Lila*—of the contemporary novelist Marilynne Robinson. Lest I sound too much of a prig, not all of my reading has had to have deep meaning—nor would Wolf and others insist that it must. (Between the lines, I sense for Wolf a fondness for Sudoku solving. I am more of a crossword man, myself.) In the realm of books, linking childhood and adulthood is Sherlock Holmes. There has never been a better short story than "The Adventure of the Speckled Band," except perhaps "The Red-Headed League" or "The Man with the Twisted Lip." Going back to meaning, as my introduction to literature was thanks to Rudyard Kipling, so I hope the last thing I ever read will be by him. If you have not done so, I beg you to read "The Gardener."

I always wanted to contribute, to be a player. As a child, in Quaker meeting when a "weighty Friend" would start pontificating, I would shut off and start planning a book. I now do the same in department meetings. From the start, I knew that I did not have the imagination of the novelist. Writing nonfiction is just as creative. Read this book! For me the play of ideas has always been vitally important and all-consuming. Fifteen minutes into my first philosophy class—it was on Descartes's *Meditations* and how we know if we are awake or asleep—I knew that *that* was what I was going to do for the rest of my life. What I did not then know was that I was going to be able to combine it with my love of history, particularly history of the Victorians. I always thought—I still think—that history on its own is great fun but not really a full-time subject for grown-ups. One needs more, and that means philosophy. I moved to philosophy of science as my special area of interest, at least in part because ethics (a natural for someone with a Quaker background) was so boring and irrelevant (no Darwinian infusion!). I had never in my life taken a course in biology—in my day, biology, like Spanish and geography, was for those known euphemistically as "late developers"—but, for the very good reason that there was not much written about it at the time, and that which was written wasn't very good, focused on it in my doctoral work. If you think of Aristotle and Kant, that surely shows that not all change is progress!

Then came Thomas Kuhn's *The Structure of Scientific Revolutions*, with its message that, if you want to understand science, you must understand the history of science. This took me straight to Darwin and the *Origin*, and here we are a half century later. Isaiah Berlin divided thinkers into two kinds, hedgehogs like Plato who saw everything through one idea, and foxes like Aristotle, who range over many ideas.[32] I am very obviously a hedgehog, but it doesn't mean that the course of one's thinking is straight down a narrow road—every turn taken, every hill ascended, shows new vistas and places to stop and try to understand. At the practical level, showing again why I am uncomfortable separating moral value from other values, my journey took me into the federal courtroom in Little Rock, Arkansas, where I was the historian and philosopher of science who spoke up for the American Civil Liberties Union, in the already-mentioned (successful) suit it brought against a law that mandated the teaching of so-called creation science (biblical literalism) along with evolutionary theorizing in biology classes in the state's public schools.[33] At the more theoretical level, Darwinism has been my lifelong passion and interest, and I am glad that I have had the chance to study such a momentous aspect of human cultural history. It has given me opportunity and inclination to always think outside the box, as it were. Wearing my hat more as a historian of science—my earlier put-down was jokingly self-referential—I have always been interested in the sociological and ideological side of things. This has led most recently to an analysis of the Darwinian revolution through my personal passion for literature, looking at the reasons why, in major respects, as I expressed above, Darwinism has always functioned as a secular religion.[34] I might regret Darwin Day, but I am not surprised by it. Wearing my hat more as a philosopher of science, it has led to thoughts about epistemology and ethics, expressed earlier in this chapter. Thoughts that, in my youth, would have made me deeply ashamed and of which—given that I am now being criticized in journals that would never accept anything by me—many today think I should still be deeply ashamed. As it happens, I have never been deeply ashamed of anything I have written—well, hardly ever!

I am finding teaching and scholarship more exciting now than ever before. I never thought I would teach a graduate course on Kant or another on American philosophy. I never thought that my love of George Eliot, Thomas Hardy, and Emily Dickinson would tell me so much about the shock of Darwinism on the Victorian mind. I never thought that, after years of making rude comments about Sewall Wright, I would now be sympathetic to a panpsychic perspective. I should add, to my surprise, I find that this move is today positively trendy in some very respectable circles. Perhaps there is some change, finally. In Canadian philosopher William Seager's excellent overview of philosophies of mind, *Theories of Consciousness*, although he finds no place in his index for either Darwin or evolution, in a discussion of the new popularity of panpsychism, he manages somewhat sheepishly to tuck in a quotation from Clifford, adding: "The addition of the theory of evolution which gives a palpable mechanism by which the simple is differentially compounded into the complex adds impetus to the slide towards a true panpsychism."[35]

Darwinism, but always with philosophy there in the end. Writing books like this, that have some interesting history but with an overall point. With a purpose, as one might say! A non-believer like me lives life day by day, finding value as one goes along. But at the end of fifty years, one can look back, as I do, with great satisfaction. Some may fault me for being elitist, stressing being a professor and so forth. To each his or her own. I certainly do not imply that what was of worth for me was of worth for all. In Dickens's great novel *David Copperfield*, the companion of David's aunt Betsy is Mr. Dick, who is feeble-minded, or however one would describe him in our politically correct society. When we first meet him, he is writing a memorial, constantly interrupted by a haunting need to refer to the lopped-off head of King Charles the First. Then Aunt Betsy loses her money and Mr. Dick, who has beautiful handwriting, turns to copying legal documents and the like, making a little money and thus supplementing their income. This is incredibly meaningful for him, even though it would not be for David, who like his creator becomes a successful writer. The story does point to, what is for me, an important part of the meaningful life, namely, striving to do better than one thought one could. All my life I have been spurred by the sense that this I must do, for I am living a life for my mother—who died suddenly at the age of thirty-three—as well as myself.

In all of this, there is a huge element of what Bernard Williams—and, to be fair, Thomas Nagel—called "moral luck."[36] In my earliest years I had very loving parents; I was raised a Quaker, and although I no longer accept the Christian God, I am so aware every day that the devotion and moral worth of the Friends of my childhood infuse every part of my being; I have been healthy; I have lived in safe times in safe places; I came to university teaching when there were still many good jobs available; I lived in Canada where the Scottish influence on higher education made for great integrity; I met Lizzie; and much more. All one can say is that I have had the chances, unlike my mother, and I have tried to use them to the full. That is a very Christian sentiment—the parable of the talents has always been binding on me—but in a way my life has been better, more purposeful, than that of the Christian or other religious believer. To mention Bernard Williams, yet again, somewhat to my surprise, eternity strikes me as a bit tedious. I have been able to find purpose in what I am doing without hope or fear of what it is worth in some ultimate sense. You cannot have a more value-impregnated—a more purpose-filled—life than that.

There is nothing more to be said. Although a closing poem is not a bad idea.

EPILOGUE

In Indianapolis they drive
five hundred miles and end up
where they started: survival
of the fittest. In the swamps
of Auburn and Elkhart,
in the jungles of South Bend,
one-cylinder chain-driven runabouts fall
to air-cooled V-4's, a-speed gearboxes,
16-horse flat-twin midships engines—
carcasses left behind
by monobloc motors, electric starters,
3-speed gears, six cylinders, 2-chain drive,
overhead cams, supercharged
to 88 miles an hour in second gear, the age
of Leviathan . . .

There is grandeur in this view of life,
as endless forms
most beautiful and wonderful
are being evolved.

And then
the drying up, the panic,
the monsters dying: Elcar, Cord,
Auburn, Duesenberg, Stutz—somewhere
out there, the chassis of Studebakers,
Marmons, Lafayettes, Bendixes, all
rusting in high-octane smog,
ashes to ashes, they
end up where they started.[1]

Is there any purpose to it all? I dunno, but it's fun while it lasts.

NOTES

Prologue

1. R. Fortey, *The Earth: An Intimate History* (New York: Vintage, 2005).

2. J. O. Farlow, C. V. Thompson, and D. E. Rosner, "Plates of the Dinosaur Stegosaurus: Forced Convection Heat Loss Fins?" *Science* 192:1123–25.

3. J. E. Lovelock, *Gaia: A New Look at Life on Earth* (Oxford: Oxford University Press, 1979).

4. M. Ruse, *The Gaia Hypothesis: Science on a Pagan Planet* (Chicago: University of Chicago Press, 2013).

5. R. Dawkins, *The Extended Phenotype: The Gene as the Unit of Selection* (Oxford: W. H. Freeman, 1982).

6. I. Kant, *Critique of the Power of Judgment*, trans. and ed. P. Guyer (Cambridge: Cambridge University Press, [1790] 2000).

7. R. J. Richards and M. Ruse, *Debating Darwin* (Chicago: University of Chicago Press, 2016).

Chapter One. Athens

1. J. Barnes, ed., *The Complete Works of Aristotle* (Princeton, NJ: Princeton University Press, 1984), 332; *Physics* 194b19.

2. M. Ruse, *Darwin and Design: Does Evolution Have a Purpose?* (Cambridge, MA: Harvard University Press, 2003).

3. R. J. Hankinson, *Cause and Explanation in Ancient Greek Thought* (Oxford: Oxford University Press, 1998), 5–6; T. Nagel, *Mind and Cosmos: Why the Materialist Neo-Darwinian Conception of Nature Is Almost Certainly False* (New York: Oxford University Press, 2012), 58–59.

4. The King James Bible version is used throughout the book for biblical references unless otherwise noted.

5. J. M. Cooper, ed., *Plato: Complete Works* (Indianapolis: Hackett, 1997), 83–84; *Phaedo*, 96 c–d.

6. Ibid.; *Phaedo*, 97 c–d.

7. Ibid.

8. D. Sedley, *Creationism and Its Critics in Antiquity* (Berkeley: University of California Press, 2007), 150–53, *De rerum natura*, 5.837–48. Copyright © 2009, The Regents of the University of California. Further refer-

ences from this source are from book 5 and appear in the main text by the line numbers.

9. Cooper, *Plato*, 289; *Sophist*, 265c.

10. Sedley, *Creationism*, 133–34. Copyright © 2009, The Regents of the University of California.

11. Ibid., 208.

12. Cooper, *Plato*, 1615; *Laws*, 967c.

13. Sedley, *Creationism*, 114. Copyright © 2009, The Regents of the University of California.

14. Cooper, *Plato*, 1235; *Timaeus*, 29a.

15. *Timaeus*, 29a.

16. Ibid., 30b.

17. Barnes, *Complete Works of Aristotle*, 688; *De Anima*, 432b22.

18. M. Leunissen, *Explanation and Teleology in Aristotle's Science of Nature* (Cambridge: Cambridge University Press, 2010).

19. J. G. Lennox, *Aristotle's Philosophy of Biology* (Cambridge: Cambridge University Press, 2001).

20. Barnes, *Complete Works of Aristotle*, 332–34; *Physics*, 194b16–195a3.

21. Ibid., 999; *Parts of Animals*, 642a32–34.

22. M. R. Johnson, *Aristotle on Teleology* (Oxford: Oxford University Press, 2005).

23. Barnes, *Complete Works of Aristotle*, 656; *De Anima*, 412a28.

24. Ibid., 661; ibid., 415b15–16.

25. Ibid., 1994–95; *Politics*, 1256b15–22.

26. Ibid., 661; *De Anima*, 415a25–415b1.

27. H. Bergson, *L'évolution créatrice* (Paris: Alcan, 1907).

28. D. H. Lawrence, *Women in Love* (London: Penguin, 1921 [1960]), 499–500.

29. Johnson, *Aristotle on Teleology*, 287.

30. Barnes, *Complete Works of Aristotle*, 342; *Physics*, 200a31-33.

31. Ibid., 340; ibid., 199a20–30.

32. Ibid.; ibid., 199a30–33.

33. Cooper, *Plato*, 444; *Philebus*, 54c.

34. *Parts of Animals*, 641b10–18. I use Sedley's 2007 translation, *Creationism and Its Critics in Antiquity*, 194, to bring out the notion of purpose.

35. Barnes, *Complete Works of Aristotle*, 389; *Physics*, 230b12–13.

36. Ibid., 1649; *Metaphysics*, 1044b8–12.

37. Ibid., 476; *De Caelo*, 289a11–12.

38. Ibid., 1694; *Metaphysics*, 1072b10–11.

39. Ibid., 1695; ibid., 1072a22–26.

40. Hankinson, *Cause and Explanation*, 187.

41. Barnes, *Complete Works of Aristotle*, 1694; *Metaphysics*, 1072a26–28.

42. Ibid., 1698; ibid., 1074b23–27.

43. Ibid.; ibid., 1074b33–34.

44. Ibid.; ibid., 1075a10–15.

45. Barnes, *Complete Works of Aristotle*; *Politics* 1253a.

46. Ibid.; *Nichomachean Ethics*, 1177a11–18.

47. Ibid.; ibid., 1177a7–9.

48. Cicero, *De Natura Deorum*, trans. H. Rackham (Cambridge, MA: Harvard University Press; Loeb Classical Library, 1933), 216–17.

49. Ibid., 246.

50. Cooper, *Plato*, 1560; *Laws*, 10.903c.

51. Barnes, *Complete Works of Aristotle*, 1802; *Nicomachean Ethics*, 1141a34–b1.

52. Cicero, *De Natura Deorum*, 158.

Chapter Two. Jerusalem

1. Augustine, *Confessions*, trans. H. Chadwick (Oxford: Oxford University Press, [396] 1998), 228.

2. Ibid., 225–26.

3. Ibid., 230.

4. Augustine, *On the Trinity*, trans. A. W. Hadden (New York: Philip Schaff, 1887), 24–25.

5. Plotinus, *The Enneads*, trans. S. MacKenna (Burdett, NY: Larson Publications, 1992), 3:14.

6. Ibid., 5:1, 5:7.

7. Augustine, *The City of God against the Pagans*, ed. and trans. R. W. Dyson (Cambridge: Cambridge University Press, [413–426] 1998), 394.

8. Ibid., 393.

9. M. Ruse, *Atheism: What Everyone Needs to Know* (Oxford: Oxford University Press, 2015).

10. Augustine, *Confessions*, 127; further references to this volume in this paragraph are given by page number.

11. Augustine, *City of God*, 53.

12. Lucretius, *Of the Nature of Things*, trans. W. E. Leonard (London: Dutton, Everyman's Library, 1950), v.

13. Augustine, *City of God*, 617.

14. Ibid., 1070.

15. J. M. Cooper, ed., *Plato: Complete Works* (Indianapolis: Hackett, 1997), 364; *Parmenides* 130d.

16. Augustine, *Confessions*, 8–9.

17. Ibid., 223.

18. Augustine, *Ad Simplicianum de diversis quaestionibus*, in *Responses to Miscellaneous Questions*, ed. B. Ramsey (Hyde Park, NY: New City Press, 2008).

19. Augustine, *The Augustine Catechism: The Enchiridion on Faith, Hope and Charity*, trans. B. Harbert (Hyde Park, NY: New City Press, 1999), 96.

20. Augustine, *Expositions of the Psalms*, trans. M. Boulding (Hyde Park, NY: New City Press, 2000), 1, 364–65.

21. Augustine, *On Free Choice of the Will* (*De Libero Arbitrio*), trans. M. Pontifex (Westminster: Newman Press, 1955), 3.4.11.

22. Augustine, *The City of God*, 510–11.

23. Ibid., 16.

24. Ibid., 544–45.

25. M. Ruse, "The Shame of Calvin College," *Brainstorm: Chronicle of Higher Education*, 2011, http://www.chronicle.com/blogs/brainstorm/the-shame-of-calvin-college/37484.

26. D. C. Lindberg, *The Beginnings of Western Science: The European Scientific Tradition in Philosophical, Religious, and Institutional Context, Prehistory to A.D. 1450* (Chicago: University of Chicago Press, 1992).

27. T. Aquinas, *Summa Theologiae*, vol. 11, *Man (1a. 75–83)* (London: Eyre and Spottiswoode, 1970), 2.

28. Augustine, *The City of God*, 413–26, 452.

29. Aquinas, *Summa Theologica*, vol. 1 (London: Burns, Oates and Washbourne, 1952), 2, 3.

30. Aquinas, *Summa Theologiae*, vol. 11, *Man (1a. 75–83)*, 2, 3.

31. Ibid.

32. O. Pedersen, *Early Physics and Astronomy: A Historical Introduction* (Cambridge: Cambridge University Press, 1993), 210.

Chapter Three. Machines

1. A. R. Hall, *The Scientific Revolution, 1500–1800: The Formation of the Modern Scientific Attitude* (London: Longman, Green, 1954), xvi–xvii.

2. M. Ruse, *Science and Spirituality: Making Room for Faith in the Age of Science* (Cambridge: Cambridge University Press, 2010).

3. R. Boyle, *A Free Enquiry into the Vulgarly Received Notion of Nature*, ed. E. B. Davis and M. Hunter (Cambridge: Cambridge University Press, 1996), 12–13.

4. R. Descartes, "Meditations," in *Philosophical Essays* (Indianapolis: Bobbs-Merrill, [1642] 1964), 111.

5. T. W. Africa, "Copernicus' Relation to Aristarchus and Pythagoras," *Isis* 52 (1961): 408, quoting *De Revolutionibus* 1.10.

6. E. A. Burtt, *The Metaphysical Foundations of Modern Science* (New York: Harcourt, Brace, 1932), 48, quoting an early fragment.

7. J. Kepler, *Harmonices Mundi* (1619), in Kepler, *The Harmony of the World*, trans. E. J. Aiton, A. M. Duncan, and J. V. Field (Philadelphia: American Philosophical Society, 1977), 358–59.

8. Ibid., 363–64.

9. Galileo, *Dialogue Concerning the Two Chief World Systems*, trans. S. Drake (New York: Random House, [1632] 2001), 11–12.

10. M. Ruse, *Darwin and Design: Does Evolution Have a Purpose?* (Cambridge, MA: Harvard University Press, 2003).

11. John Keble, "Who Runs May Read," in *The Christian Year*, published 1827.

12. H. More, *The Immortality of the Soul* (Dordrecht: Nijhoff, [1659] 1987), lxxxi.

13. A. R. Hall, *Henry More: Magic, Religion and Experiment* (Oxford: Blackwell, 1990).

14. Kepler, in a 1605 letter to Herwart von Hohenburg, cited in G. Holton, *Thematic Origins of Scientific Thought* (Cambridge, MA: Harvard University Press, 1973), 72.

15. E. J. Dijksterhuis, *The Mechanization of the World Picture* (Oxford: Oxford University Press, 1961), 491.

16. R. Boyle, "A Disquisition about the Final Causes of Natural Things," in *The Works of Robert Boyle*, ed. T. Birch (Hildesheim: Georg Olms, [1688] 1966), 5:397–98.

17. Ibid.

18. Ibid., 5:424.

19. Ibid., 5:428.

20. D. Hume, *Dialogues Concerning Natural Religion*, ed. M. Bell (London: Penguin, [1779] 1990), 77.

21. Ibid., 108–9.

22. Ibid., 109.

23. Ibid., 130.

24. P. Lipton, *Inference to the Best Explanation* (London: Routledge, 1991).

25. W. Paley, *Natural Theology* (*Collected Works: Volume 4*) (London: Rivington, [1802] 1819), 1.

26. Ibid., 13–14.

27. Cecil Frances Alexander, third stanza from "All Things Bright and Beautiful," in *Hymns for Little Children*, 1848.

28. D. Hume, *A Treatise of Human Nature*, ed. D. F. Norton and M. J. Norton (Oxford: Oxford University Press, [1739–40] 2000), 302.

29. I. Kant, *Critique of Pure Reason*, trans. and ed. P. Guyer and A. W. Wood (Cambridge: Cambridge University Press [1781/1787] 1998), 117.

30. I. Kant, *Religion within the Bounds of Mere Reason*, ed. A. Wood and G. di Giovanni (Cambridge: Cambridge University Press, [1793] 1998), 36.

31. B. Spinoza, "Ethics," in *The Collected Writings of Spinoza*, trans. E. Curley (Princeton, NJ: Princeton University Press, [1677] 1985).

32. I. Kant, *Critique of the Power of Judgment*, ed. P. Guyer, trans. E. Matthews (Cambridge: Cambridge University Press, [1790] 2000), 246.

33. Ibid., 247–48; boldface in original.

34. Ibid., 246.

35. G.W.F. Hegel, *Logic*, trans. W. Wallace (Pacifica, CA: Marxist's Internet Archive, [1830] 2008), 442.

36. Kant, *Critique of the Power of Judgment*, 247.

37. Ibid., 271.

38. M. R. Johnson, *Aristotle on Teleology* (Oxford: Oxford University Press, 2005), 188.

Chapter Four. Evolution

1. M. Ruse, *Monad to Man: The Concept of Progress in Evolutionary Biology* (Cambridge, MA: Harvard University Press, 1996).

2. J. B. Bury, *The Idea of Progress: An Inquiry into Its Origin and Growth* (London: Macmillan, [1920] 1924), 5.

3. I. Kant, *Critique of the Power of Judgment*, trans. and ed. P. Guyer (Cambridge: Cambridge University Press, [1790] 2000), 302–3.

4. M. Ruse, *The Evolution-Creation Struggle* (Cambridge, MA: Harvard University Press, 2005).

5. W. Godwin, *An Enquiry Concerning Political Justice* (Oxford: Oxford University Press, [1793] 2013), book 1, chapter 5.

6. Ibid., book 2, chapter 4.

7. Adam Smith, *An Inquiry into the Nature and Causes of the Wealth of Nations*, book 1, chapter 2, "Of the Principle which gives occasion to the Division of Labour" (1776).

8. D. Diderot, *Diderot: Interpreter of Nature* (New York: International, 1943), 152.

9. G. M. Kirk, J. E. Raven, and M. Schofield, *The Presocratic Philosophers: A Critical History with a Selection of Texts* (Cambridge: Cambridge University Press, 1984), 303.

10. P. J. Bowler, *Evolution: The History of an Idea* (Berkeley: University of California Press, 1984).

11. Kant, *Critique of the Power of Judgment* (1790), 288.

12. D. Cuvier, *Le règne animal distribué d'aprés son organisation, pour servir de base à l'histoire naturelle des animaux et d'introduction à l'anatomie compare* (Paris, 1817), 1, 3–4.

13. Kant, in *Critique of the Power of Judgment* (1790), 287.

14. Diderot, *Diderot*, 48.

15. E. Darwin, *The Temple of Nature* (London: J. Johnson, 1803), 1, 11, 295–314.

16. E. Darwin, *Zoonomia; or, The Laws of Organic Life*, 3rd ed. (London: J. Johnson, [1794–96] 1801), 2:247–48.

17. R. Chambers, *Vestiges of the Natural History of Creation* (London: J. Churchill, 1844), 400–402.

18. Ibid.

19. Ibid.

20. M. Ruse, *Monad to Man*, 1996.

21. R. W. Burkhardt, *The Spirit of System: Lamarck and Evolutionary Biology* (Cambridge, MA: Harvard University Press, 1977).

22. W. Whewell, *The Philosophy of the Inductive Sciences* (London: Parker, 1840).

23. Kant, *Critique of the Power of Judgment* (1790), 245.

24. R. Owen, *On the Archetype and Homologies of the Vertebrate Skeleton* (London: Voorst, 1848).

25. Ibid.

26. C. C. Gillespie, *Genesis and Geology* (Cambridge, MA: Harvard University Press, 1950).

27. W. Whewell, *Astronomy and General Physics* (*Bridgewater Treatise, 3*) (London: William Pickering, 1833).

28. W. Whewell, *Of the Plurality of Worlds* (London: Parker, 1853), 221. This later work took up the plurality problem explicitly and at length. It is from this work, or rather my facsimile edition (Chicago: University of Chicago Press, 2001), that the quotations are taken. The work was published anonymously because by this time Whewell was Master of Trinity College, Cambridge, and open controversy would be considered unseemly. Everyone knew the identity of the author.

29. Ibid., 222

30. Ibid., 226.

31. M. Ruse, "Kant and Evolution," in *Theories of Generation*, ed. J. Smith, 402–15 (Cambridge: University of Cambridge Press, 2006).

32. Kant, *Critique of the Power of Judgment* (1790), 288.

33. P. H. Barrett et al., *Charles Darwin's Notebooks, 1836–1844* (Ithaca, NY: Cornell University Press, 1987), 174.

34. W. Whewell, *The History of the Inductive Sciences* (London: Parker, 1837), 3, 588.

Chapter Five. Charles Darwin

1. J. Browne, *Charles Darwin: Voyaging, Volume 1 of a Biography* (New York: Knopf, 1995), and *Charles Darwin: The Power of Place, Volume 2 of a Biography* (New York: Knopf, 2002).

2. R. J. Richards and M. Ruse, *Debating Darwin* (Chicago: University of Chicago Press, 2016).

3. M. Ruse, *Darwinism as Religion: What Literature Tells Us about Evolution* (Oxford: Oxford University Press, 2017).

4. M. Ruse, "Darwin's Debt to Philosophy: An Examination of the Influence of the Philosophical Ideas of John F. W. Herschel and William Whewell on the Development of Charles Darwin's Theory of Evolution," *Studies in History and Philosophy of Science* 6 (1975): 159–81.

5. T. R. Malthus, *An Essay on the Principle of Population* (New York: Macmillan, [1798] 1966).

6. C. Darwin, *On the Origin of Species by Means of Natural Selection, or the Preservation of Favoured Races in the Struggle for Life* (London: John Murray, 1859), 63–64.

7. T. Kuhn, *The Structure of Scientific Revolutions* (Chicago: University of Chicago Press, 1962).

8. C. Darwin, *On the Origin of Species*, 80–81.

9. W. Whewell, *The Philosophy of the Inductive Sciences* (London: Parker, 1840).

10. M. Ruse, "Sexual Selection: Why Does It Play Such a Large Role in the *Descent of Man?*," in *Current Perspectives on Sexual Selection: What's Left after Darwin?*, ed. T. Hoquet (New York: Springer, 2015), 3–17.

11. C. Darwin, *The Autobiography of Charles Darwin, 1809–1882*, ed. Nora Barlow (London: Collins, 1958).

12. W. Henry, *The Elements of Experimental Chemistry*, 8th ed. (London: Baldwin, Cradock and Joy, 1818), xix.

13. Ibid., iii.

14. W. Kirby and W. Spence, *An Introduction to Entomology: or, Elements of the Natural History of Insects* (London: Longman, Hurst, Reece, Orme, and Brown, 1815–28).

15. Ibid., xvi.

16. Cuvier, *Le règne animal distribué d'après son organisation pour servir de base a l'histoire naturelle des animaux* (1817), 1, 6.

17. Darwin, *On the Origin of Species* (1859), 469.

18. Darwin, *Origin of Species*, 3rd ed. (1861), 163.

19. A. Desmond, *Huxley: From Devil's Disciple to Evolution's High Priest* (New York: Basic Books, 1997).

20. Darwin, *On the Origin of Species* (1859), 206.

21. R. Dawkins, *The Blind Watchmaker* (New York: Norton, 1986). Alvin Plantinga, in his *Warrant and Proper Function*, notes correctly that Kant offered a heuristic solution without explaining why it works. Plantinga argues that this shows we cannot have a natural explanation of organic purpose or function, a conclusion he is able to achieve only by ignoring entirely the Darwinian solution, almost too typical a gambit by analytic philosophers to be worthy of remark.

22. R. Dawkins, *The Selfish Gene* (Oxford: Oxford University Press, 1976), 21.

23. C. Darwin and A. R. Wallace, *Evolution by Natural Selection*, foreword by Gavin de Beer (Cambridge: Cambridge University Press, 1958), 45–46.

24. Darwin, *On the Origin of Species* (1859), 488.

25. C. Darwin, *The Correspondence of Charles Darwin* (Cambridge: Cambridge University Press, 1985–), 8, 224; May 22, 1860.

26. A. Gotthelf, "Darwin on Aristotle," *Journal of the History of Biology* 32:3–30.

27. Darwin, *Origin of Species*, 3rd ed. (1861), 85.

28. C. Darwin, *On the Various Contrivances by which British and Foreign Orchids Are Fertilized by Insects, and On the Good Effects of Intercrossing* (London: John Murray, 1862).

29. Dawkins, *The Blind Watchmaker*, 6.

30. Darwin, *On the Origin of Species* (1859), 490.

31. P. H. Barrett et al., *Charles Darwin's Notebooks, 1836–1844* (Ithaca, NY: Cornell University Press, 1987), Notebook E, 95–96.

32. Darwin, *Origin of Species*, 3rd ed. (1861), 133.

33. Ibid., 134.

Chapter Six. Darwinism

1. C. Naden, *Poetical Works of Constance Naden* (Kernville, CA: High Sierra Books, 1999), 207–8; written around 1885.

2. M. Ruse, *Monad to Man: The Concept of Progress in Evolutionary Biology* (Cambridge, MA: Harvard University Press, 1996).

3. W. B. Provine, *The Origins of Theoretical Population Genetics* (Chicago: University of Chicago Press, 1971).

4. P. R. Grant, *Ecology and Evolution of Darwin's Finches* (Princeton, NJ: Princeton University Press, 1986); P. R. Grant and R. B. Grant, *How and Why Species Multiply: The Radiation of Darwin's Finches* (Princeton, NJ: Princeton University Press, 2007).

5. J. A. Hopson, "The Evolution of Cranial Display Structures in Hadrosaurian Dinosaurs," *Paleobiology* 1 (1975): 21–43; D. B. Weishampel, "Acoustic Analyses of Potential Vocalization in Lambeosaurine Dinosaurs (Reptilia: Ornithischia)," *Paleobiology* 7 (1981): 252–61.

6. D. B. Weishampel, "Dinosaurian Cacophony," *BioScience* 47, no. 3 (1997): 150–58.

7. M. Ruse, *Darwinism Defended: A Guide to the Evolution Controversies* (Reading, MA: Benjamin/Cummings, 1982).

8. T. Dobzhansky, F. J. Ayala, G. L. Stebbins, and J. W. Valentine, *Evolution* (San Francisco: Freeman, 1977).

9. R. C. Lewontin, *The Genetic Basis of Evolutionary Change* (New York: Columbia University Press, 1974).

10. S. J. Gould, *Ontogeny and Phylogeny* (Cambridge, MA: Belknap Press, 1977) and *The Structure of Evolutionary Theory* (Cambridge, MA: Harvard University Press, 2002).

11. S. J. Gould and R. C. Lewontin, "The Spandrels of San Marco and the Panglossian Paradigm: A Critique of the Adaptationist Programme," *Proceedings of the Royal Society of London, Series B: Biological Sciences* 205 (1979): 581–98.

12. Ibid., 582.

13. J. M. Smith, "Did Darwin Get It Right?," *London Review of Books* 3, no. 11 (1981): 10–11.

14. S. Wright, "Evolution in Mendelian Populations," *Genetics* 16 (1931): 97–159 and "The Roles of Mutation, Inbreeding, Crossbreeding and Selection in Evolution," *Proceedings of the Sixth International Congress of Genetics* 1 (1932): 356–66.

15. J. A. Coyne, N. H. Barton, and M. Turelli, "Perspective: A Critique of Sewall Wright's Shifting Balance Theory of Evolution," *Evolution* 51, no. 3 (1997): 643–71.

16. Smith, "Did Darwin Get It Right?," 10–11.

17. M. Ruse, *Darwin and Design: Does Evolution Have a Purpose?* (Cambridge, MA: Harvard University Press, 2003).

18. C. Allen, M. Bekoff, and G. Lauder, *Nature's Purposes, Analyses of Function and Design in Biology* (Cambridge, MA: MIT Press, 1998).

19. L. Wright, "Functions," *Philosophical Review* 82, no. 2 (April 1973): 139–68.

20. R. Cummins, "Functional Analysis," *Journal of Philosophy* 72, no. 20 (Nov. 20, 1975): 741–65.

21. M. Ruse, *The Philosophy of Biology* (London: Hutchinson, 1973).

22. M. Mossio, C. Saborido, A. Moreno, "An Organizational Account of Biological Functions," *British Journal for the Philosophy of Science* 60, no. 4 (2009): 813–41.

23. M. Bedau, "Where's the Good in Teleology?," *Philosophy and Phenomenological Research* 52 (1992): 781n1.

24. F. Ayala, "Teleological Explanations in Evolutionary Biology," *Philosophy of Science* 37 (1970): 1–15.

25. E. Nagel, *The Structure of Science: Problems in the Logic of Scientific Explanation* (New York: Harcourt, Brace and World, 1961).

26. S. J. Gould, "On Replacing the Idea of Progress with an Operational Notion of Directionality," in *Evolutionary Progress*, ed. M. H. Nitecki (Chicago: University of Chicago Press, 1988), 319.

27. S. J. Gould, *Wonderful Life: The Burgess Shale and the Nature of History* (New York: W. W. Norton, 1989), 318.

28. E. O. Wilson, *The Diversity of Life* (Cambridge, MA: Harvard University Press, 1992), 187.

29. E. O. Wilson, *Sociobiology: The New Synthesis* (Cambridge, MA: Harvard University Press, 1975).

30. G. C. Williams, *Adaptation and Natural Selection* (Princeton, NJ: Princeton University Press, 1966).

31. R. Dawkins, "Progress," in *Keywords in Evolutionary Biology*, ed. E. F. Keller and E. Lloyd (Cambridge, MA: Harvard University Press, 1992), 265–66.

32. M. Ruse, *The Philosophy of Biology*.

33. C. Darwin, *On the Origin of Species by Means of Natural Selection, or the Preservation of Favoured Races in the Struggle for Life* (London: John Murray, 1859), 345.

34. Jack Sepkoski, interview with the author, January 1989.

35. Keith Stewart Thomson, "The Pattern of Diversification among Fishes," in *Patterns of Evolution as Illustrated by the Fossil Record*, edited by A. Hallam (Amsterdam: Elsevier, 1977), 5:547–62.

36. J. A. Doyle, "Patterns of Evolution in Early Angiosperms," in *Patterns of Evolution as Illustrated by the Fossil Record*, ed. A. Hallam (Amsterdam: Elsevier, 1977), 5:531.

37. J. S. Huxley, *The Individual in the Animal Kingdom* (Cambridge:

Cambridge University Press, 1912), 115–16; all subsequent quotes in this paragraph are from these page numbers in this source.

38. R. Dawkins and J. R. Krebs, "Arms Races Between and Within Species," *Proceedings of the Royal Society of London, B* 205:508.

39. H. Jerison, *Evolution of the Brain and Intelligence* (New York: Academic Press, 1973).

40. R. Dawkins, *The Blind Watchmaker* (New York: Norton, 1986), 189.

41. Conference talk, Melbu, Norway, 1989; printed in Ruse, *Monad to Man*, 469.

42. Ibid.

43. Ruse, *Monad to Man*, 486.

44. S. J. Gould, *The Flamingo's Smile: Reflections in Natural History* (New York: Norton, 1985), 412.

45. Ibid.

46. S. Conway Morris, *Life's Solution: Inevitable Humans in a Lonely Universe* (Cambridge: Cambridge University Press, 2003).

47. Ibid., 196.

48. J. S. Huxley, *Evolution: The Modern Synthesis* (London: Allen and Unwin, 1942).

49. J.B.S. Haldane, *Possible Worlds and Other Essays* (London: Chatto and Windus, 1927), 286.

Chapter Seven. Plato Redivivus

1. W. Whewell, *The History of the Inductive Sciences* (London: Parker, 1837); A. A. Sedgwick, "Address to the Geological Society," *Proceedings of the Geological Society of London* 1 (1831): 281–316.

2. E. Lurie, *Louis Agassiz: A Life in Science* (Chicago: University of Chicago Press, 1960).

3. A. Gray, review of *The Origin of Species by Means of Natural Selection*, *American Journal of Arts and Sciences*, in *Darwiniana* (New York: Appleton, [1860] 1876), 7.

4. C. Darwin, *The Variation of Animals and Plants under Domestication* (London: Murray, 1868), 2, 432.

5. A. R. Wallace, "The Limits of Natural Selection as Applied to Man," in *Contributions to the Theory of Natural Selection* (London: Macmillan, [1865] 1870), 359–60.

6. Augustine, *The City of God against the Pagans*, ed. and trans. R. W. Dyson (Cambridge: Cambridge University Press, [413–426] 1998), 512.

7. M. Noll, *America's God: From Jonathan Edwards to Abraham Lincoln* (New York: Oxford University Press, 2002).

8. R. L. Numbers, *The Creationists: From Scientific Creationism to Intelligent Design* (Cambridge, MA: Harvard University Press, 2006).

9. J. C. Whitcomb and H. M. Morris, *The Genesis Flood: The Biblical Record and Its Scientific Implications* (Philadelphia: Presbyterian and Reformed Publishing Company, 1961).

10. H. M. Morris, "Design Is Not Enough," *Back to Genesis 127* (1999): a–c.

11. M. Ruse, *But Is It Science? The Philosophical Question in the Creation/Evolution Controversy* (Buffalo, NY: Prometheus, 1988).

12. P. E. Johnson, *Darwin on Trial* (Washington, DC: Regnery Gateway, 1991).

13. M. Behe, *Darwin's Black Box: The Biochemical Challenge to Evolution* (New York: Free Press, 1996), 70.

14. K. Miller, *Finding Darwin's God* (New York: Harper and Row, 1999).

15. F. Jacob, "Evolution and Tinkering," *Science* 196 (1977): 1161–66.

16. R. F. Doolittle, "A Delicate Balance," *Boston Review* 22, no. 1 (1997): 28–29.

17. John Paul II, "The Pope's Message on Evolution," *Quarterly Review of Biology* 72 (1997): 377–83.

18. R. J. Russell, *Cosmology: From Alpha to Omega: The Creative Mutual Interaction of Theology and Science* (Minneapolis: Fortress Press, 2008).

19. E. Sober, "Evolutionary Theory, Causal Completeness, and Theism: The Case of 'Guided' Mutations," in *Evolutionary Biology: Conceptual, Ethical, and Religious Issues*, ed. R. P. Thompson and D. M. Walsh (Cambridge: Cambridge University Press, 2014), 31–44.

20. M. Ruse, *The Gaia Hypothesis: Science on a Pagan Planet* (Chicago: University of Chicago Press, 2013).

21. F. Doolittle, "Is Nature Really Motherly?," *CoEvolution* 29 (1981): 58–62.

22. J. D. Barrow and F. J. Tipler, *The Anthropic Cosmological Principle* (Oxford: Clarendon Press, 1986).

23. Ibid., 16.

24. Ibid., 22.

25. Steven Weinberg, "A Designer Universe?," *New York Review of Books*, October 21, 1999, 47.

26. V. J. Stenger, *The Fallacy of Fine-Tuning: Why the Universe Is Not Designed for Us* (Buffalo, NY: Prometheus, 2011), 70

27. B. C. Jantzen, *An Introduction to Design Arguments* (Cambridge: Cambridge University Press, 2014).

28. G.W.F. Hegel, *Logic*, trans. W. Wallace (Pacifica, CA: Marxist's Internet Archive, [1830] 2008), 444.

29. S. Kierkegaard, *Fear and Trembling*, ed. and trans. A. Hannay (London: Penguin, [1843] 1985).

30. K. Barth, *The Epistle to the Romans* (Oxford: Oxford University Press, 1933), 134.

Chapter Eight. Aristotle Redivivus

1. Galileo, *Dialogue Concerning the Two Chief World Systems*, trans. S. Drake (New York: Random House, [1632] 2001); selections from the *Third Day*, 153–62, 165–67.

2. A. M. Smith, "Descartes's Theory of Light and Refraction: A Discourse on Method," *Transactions of the American Philosophical Society* (1987) 77:16–17; this is from the "Treatise on Light."

3. Pierre Louis Moreau de Maupertuis, "Accord between Different Laws of Nature that Seemed Incompatible," paper submitted to the Paris Academy, 1744.

4. I. Kant, *Metaphysical Foundations of Natural Science*, trans. M. Friedman (Cambridge: Cambridge University Press, [1786] 2004), 30.

5. B. Spinoza, "Ethics," in *The Collected Writings of Spinoza*, trans. E. Curley (Princeton, NJ: Princeton University Press, [1677] 1985).

6. R. J. Richards, *The Romantic Conception of Life: Science and Philosophy in the Age of Goethe* (Chicago: University of Chicago Press, 2002).

7. F.W.J. Schelling, *Ideas for a Philosophy of Nature as Introduction to the Study of This Science, 1797*, 2nd ed., trans. E. E. Harris and P. Heath (Cambridge: Cambridge University Press, [1803] 1988), 54.

8. Ibid., 35.

9. Ibid.

10. Ibid.

11. Ibid., 42.

12. D. W. Thompson, *On Growth and Form*, 2nd ed. (Cambridge: Cambridge University Press, 1948); the first edition was published in 1917.

13. S. J. Gould, "D'Arcy Thompson and the Science of Form," *New Literary History* 2 (1971): 229–58.

14. Thompson, *On Growth and Form*, 10.

15. Ibid., 395.

16. Ibid., 395–96.

17. Ibid., 966.

18. Ibid.

19. S. A. Kauffman, *At Home in the Universe: The Search for the Laws of Self-Organization and Complexity* (New York: Oxford University Press, 1995).

20. A. Gray, *Structural Botany*, 6th ed. (London: Macmillan, 1881).

21. B. Goodwin, *How the Leopard Changed Its Spots*, 2nd ed. (Princeton, NJ: Princeton University Press, 2001), 127.

22. D. King, "An Interview with Professor Brian Goodwin," *GenEthics News* 11 (1996): 6–8; quotes in the remainder of this paragraph are to this interview.

23. J.-B. Lamarck, *Philosophie zoologique* (Paris: Dentu, 1809).

24. M. Ruse, *The Darwinian Revolution: Science Red in Tooth and Claw* (Chicago: University of Chicago Press, 1979).

25. G.W.F. Hegel, *Philosophy of Nature* (Oxford: Oxford University Press, [1817] 1970), 20.

26. H. Spencer, *Autobiography* (London: Williams and Norgate, 1904).

27. H. Spencer, "Progress: Its Law and Cause," *Westminster Review* 67 (1857): 2–3.

28. H. Spencer, *First Principles* (London: Williams and Norgate, 1862).

29. H. Spencer, "The Social Organism," *Westminster Review* (1860).

30. H. Spencer, *The Data of Ethics* (London: Williams and Norgate, 1879), 21.

31. Ibid.

32. H. Driesch, *The Science and the Philosophy of the Organism* (London: Adam and Charles Black, 1908).

33. H. Bergson, *Creative Evolution* (New York: Holt, 1911), 43.

34. Ibid., 182.

35. Ibid., 185.

36. D. H. Lawrence, *Women in Love* (London: Penguin, [1921] 1960), 538.

37. J. S. Huxley, *Evolution: The Modern Synthesis* (London: Allen and Unwin, 1942).

38. Pierre Teilhard de Chardin, *Le phénomène humain* (Paris: Editions de Seuil, 1959).

39. P. B. Medawar, "Review of *The Phenomenon of Man*," *Mind* 70 (1961): 99–100.

40. S. J. Gould, "The Piltdown Conspiracy," *Natural History* 89 (August 1980): 8–28.

41. A. N. Whitehead and B. Russell, *Principia Mathematica*, 3 vols. (Cambridge: Cambridge University Press, 1910–13).

42. A. N. Whitehead, *Science and the Modern World* (New York: Free Press, [1925] 1967), 107.

43. S. Alexander, *Space, Time and Deity (The Gifford Lectures at Glasgow, 1916–1918)* (London: Macmillan, 1920).

44. Whitehead, *Science and the Modern World*, 94.

45. Ibid., 84.

46. P. Rogers, *Song of the World Becoming: New and Collected Poems, 1981–2001* (Minneapolis: Milkweed, 2001).

47. A. N. Whitehead, *Adventures of Ideas* (New York: Macmillan, 1933), 166.

48. A. N. Whitehead, *Process and Reality: An Essay in Cosmology* (New York: Free Press, [1929] 1978), 47, 343, 344.

49. M. Dibben, "Exploring the Processual Nature of Trust and Cooperation in Organisations: A Whiteheadian Analysis," *Philosophy of Management* 4 (2004): 25–39.

50. W. B. Provine, *Sewall Wright and Evolutionary Biology* (Chicago: University of Chicago Press, 1986); M. Ruse, *Monad to Man: The Concept of Progress in Evolutionary Biology* (Cambridge, MA: Harvard University Press, 1996).

51. S. Wright, "Evolution in Mendelian Populations," *Genetics* 16 (1931): 155; S. Wright, *Evolution, Selected Papers*, ed. W. B. Provine (Chicago: University of Chicago Press, 1986).

52. J. S. Huxley, *The Individual in the Animal Kingdom* (Cambridge: Cambridge University Press, 1912), 116.

53. Wright, "Evolution in Mendelian Populations," 111.

54. M. Ruse, *The Gaia Hypothesis: Science on a Pagan Planet* (Chicago: University of Chicago Press, 2013).

55. S. J. Gould, *Full House: The Spread of Excellence from Plato to Darwin* (New York: Paragon, 1996).

56. D. McShea and R. Brandon, *Biology's First Law: The Tendency for Diversity and Complexity to Increase in Evolutionary Systems* (Chicago: University of Chicago Press, 2010).

57. Ibid., 4.

58. Ibid., 124.

59. Ibid.

60. Ibid.

61. Ibid., 134.

62. Aristotle, *The History of Animals*, trans. D. W. Thompson, in *The Complete Works of Aristotle*, ed. J. Barnes (Princeton, NJ: Princeton University Press, 1984), 774–993.

63. R. Wright, "Evolution and Higher Purpose," meaningoflife.tv, 2016, http://meaningoflife.tv/articles/wright-evolution-purpose.

64. J. E. Lovelock, *Gaia: A New Look at Life on Earth* (Oxford: Oxford University Press, 1979).

Chapter Nine. Human Evolution

1. M. Ruse, *The Philosophy of Human Evolution* (Cambridge: Cambridge University Press, 2012).

2. C. Darwin, *The Descent of Man, and Selection in Relation to Sex* (London: John Murray, 1871), 1, 10.

3. P. Shipman, *The Man Who Found the Missing Link: Eugene Dubois and His Lifelong Quest to Prove Darwin Right* (Cambridge, MA: Harvard University Press, 2002).

4. D. Falk, *The Fossil Chronicles: How Two Controversial Discoveries Changed Our View of Human Evolution* (Berkeley: University of California Press, 2012).

5. D. Johanson and M. Edey, *Lucy: The Beginnings of Humankind* (New York: Simon and Schuster, 1981).

6. M. Morwood and P. Van Oosterzee. *A New Human: The Startling Discovery and Strange Story of the "Hobbits" of Flores, Indonesia* (London: Collins, 2007).

7. M. Kimura, *The Neutral Theory of Molecular Evolution* (Cambridge: Cambridge University Press, 1983).

8. M. Krings et al., "Neanderthal DNA Sequences and the Origin of Modern Humans," *Cell* 90 (1997): 19–30; J. P. Noonan et al., "Sequencing and Analysis of Neanderthal Genomic DNA," *Science* 314 (2012): 1113–18.

9. G. C. Conroy and H. Pontzer, *Reconstructing Human Origins: A Modern Synthesis*, 3rd ed. (New York: W. W. Norton, 2012).

10. T. D. White et al., "*Ardipithecus ramidus* and the Paleobiology of Early Hominids," *Science* 326, 5949 (2009): 75–86.

11. D. Falk, *Braindance: New Discoveries about Human Origins and Brain Evolution* (Gainesville: University of Florida Press, 2004).

12. C. Stringer, "Modern Human Origins: Progress and Prospects," *Philosophical Transactions of the Royal Society, London (B)* 357 (2002): 563–79; and "Human Evolution: Out of Ethiopia," *Nature* 423 (2003): 692–95.

13. R. W. El-Sabaawi et al., "Assessing the Effects of Guppy Life History Evolution on Nutrient Recycling: From Experiments to the Field," *Freshwater Biology* 60 (2015): 590–601.

14. D. N. Reznick, *The "Origin" Then and Now: An Interpretive Guide to the "Origin of Species"* (Princeton, NJ: Princeton University Press, 2009).

15. D. N. Reznick, "Guppies and the Empirical Study of Adaptation," in *In Light of Evolution: Essays from the Laboratory and Field*, ed. J. B. Losos (Greenwood Village, CO: Roberts, 2011), 205–32.

16. S. J. Gould, *Ontogeny and Phylogeny* (Cambridge, MA: Belknap Press, 1977), 504.

17. S. J. Gould, *The Mismeasure of Man* (New York: Norton, 1981).

18. V. Reynolds and R. Tanner, *The Biology of Religion* (London: Longman, 1983).

19. J. A. Coyne, *Why Evolution is True* (New York: Viking, 2009).

20. M. Dixon and G. Radick, *Darwin in Ilkley* (Stroud, Gloucestershire: History Press, 2009).

Chapter Ten. Mind

1. M. Ruse, *Philosophy after Darwin* (Princeton, NJ: Princeton University Press, 2009).

2. W. James, "Great Men, Great Thoughts, and the Environment," *Atlantic Monthly* 46, 276 (1880): 441–59.

3. S. Cunningham, *Philosophy and the Darwinian Legacy* (Rochester: University of Rochester Press, 1996).

4. H. Sidgwick, "The Theory of Evolution in Its Application to Practice," *Mind* 1 (1876): 52–67.

5. G. E. Moore, *Principia Ethica* (Cambridge: Cambridge University Press), 34.

6. B. Russell, *Power: A New Social Analysis* (London: Allen and Unwin, 1938).

7. B. Russell, *Our Knowledge of the External World as a Field for Scientific Method in Philosophy* (Chicago: Open Court, 1914), 15.

8. R. Rhees, ed., *Ludwig Wittgenstein: Personal Recollections* (Oxford: Blackwell, 1981), 174.

9. L. Wittgenstein, *Tractatus Logico-Philosophicus* (London: Routledge & Kegan Paul, 1922), 4.1122.

10. M. Ruse, *Monad to Man: The Concept of Progress in Evolutionary Biology* (Cambridge, MA: Harvard University Press, 1996).

11. M. Ruse, *The Evolution-Creation Struggle* (Cambridge, MA: Harvard University Press, 2005).

12. T. Baldwin, *G. E. Moore* (London: Routledge and Kegan Paul, 1990), 50.

13. B. Russell, *My Philosophical Development* (London: Allen and Unwin, 1959), 155.

14. W.V.O. Quine, *Ontological Relativity and Other Essays* (New York: Columbia University Press, 1969); John Rawls, *A Theory of Justice* (Cambridge, MA: Harvard University Press, 1971).

15. J. Fodor and M. Piattelli-Palmarini, *What Darwin Got Wrong* (New York: Farrar, Straus and Giroux, 2010).

16. P. R. Grant and R. B. Grant, *How and Why Species Multiply: The Radiation of Darwin's Finches* (Princeton, NJ: Princeton University Press, 2007); J. A. Coyne and H. A. Orr, *Speciation* (Sunderland, MA: Sinauer, 2004); D. Reznick, "Guppies and the Empirical Study of Adaptation," in *In Light of Evolution: Essays from the Laboratory and Field*, ed. J. B. Losos (Greenwood Village, CO: Roberts, 2011); D. Johanson and M. Edey, *Lucy: The Beginnings of Humankind* (New York: Simon and Schuster, 1981).

17. S. B. Carroll, *Endless Forms Most Beautiful: The New Science of Evo Devo* (New York: Norton, 2005); quote is from Fodor and Piattelli-Palmarini, *What Darwin Got Wrong*, 32.

18. Fodor and Piattelli-Palmarini, *What Darwin Got Wrong*, 166. At least these authors are convinced of the truth of evolution, which is more than one can say of Alvin Plantinga. See his "Where Faith and Reason Clash: Evolution and the Bible," *Christian Scholars Review*, 21 (1991), 8–32.

19. R. Hursthouse, *On Virtue Ethics* (Oxford: Oxford University Press, 1999).

20. T. Nagel, *Mind and Cosmos: Why the Materialist Neo-Darwinian Conception of Nature Is Almost Certainly False* (New York: Oxford University Press, 2012), 5.

21. Ibid., 6.

22. Ibid., 66.

23. T. H. Huxley, "On the Hypothesis that Animals Are Automata, and Its History," *Fortnightly Review* 16 (1874): 555–80.

24. W. James, *The Principles of Psychology* (New York: Henry Holt, 1880), 1, 138.

25. S. Pinker, *How the Mind Works* (New York: Norton, 1997).

26. D. Hume, *An Enquiry Concerning Human Understanding* (Oxford: Oxford University Press, [1748] 2007), 76.

27. Ibid.

28. S. Mithen, *The Prehistory of the Mind* (London: Thames and Hudson, 1996).

29. N. Chomsky, *Syntactic Structures* (The Hague: Mouton, 1957).

30. S. Pinker, *The Language Instinct: How the Mind Creates Language* (New York: William Morrow, 1994).

31. C. Holden, "The Origin of Speech," *Science* 303 (2004): 1316–19.

32. P. Lieberman, *The Biology and Evolution of Language* (Cambridge, MA: Harvard University Press, 1984).

33. B. Arensburg et al., "A Middle Paleolithic Human Hyoid Bone," *Nature* 338 (1989): 758–60.

34. J. Barnes, ed., *The Complete Works of Aristotle* (Princeton, NJ: Princeton University Press, 1984), 335; *Physics* 196b22–23.

35. T. Nagel, *Mortal Questions* (Cambridge: Cambridge University Press, 1979), 146.

36. T. Nagel, *Mind and Cosmos*, 45.

37. Ibid., 4–5.

38. Ibid., 7.

39. Ibid., 91.

40. Ibid., 120.

41. J. L. Bada and A. Lazcana, "The Origin of Life," in *Evolution: The First Four Billion Years*, ed. M. Ruse and J. Travis (Cambridge, MA: Harvard University Press, 2009), 49–79.

42. D. J. Chalmers, *The Conscious Mind* (New York: Oxford University Press, 1996).

43. J. Kim, *Supervenience and Mind: Selected Philosophical Essays* (Cambridge: Cambridge University Press, 1993), 167.

44. G. Strawson et al., *Consciousness and Its Place in Nature: Does Physicalism Entail Panpsychism?* (Exeter: Imprint Academic, 2006).

45. M. Pollan, "The Intelligent Plant: Scientists Debate a New Way of Understanding Flora," *New Yorker* (December 2013): 92–105.

46. D. Skrbina, *Panpsychism in the West* (Cambridge, MA: MIT Press, 2005), 146.

47. E. Haeckel, "Our Monism: The Principles of a Consistent, Unitary World-View," *Monist* 2 (1892): 486.

48. W. Seager, *Theories of Consciousness: An Introduction and Assessment*, 2nd ed. (London: Routledge, 2016), 297.

49. M. Lockwood, *Mind, Brain and the Quantum: The Compound "I"* (Oxford: Blackwell, 1989); H. Atmanspacher, "Quantum Approaches to Consciousness," 2015, *Stanford Encyclopedia of Philosophy*, https://plato.stanford.edu/entries/qt-consciousness/#4.1.

50. Pinker, *How the Mind Works*.

51. Nagel, *Mortal Questions*.

52. Nagel, *Mind and Cosmos*, 14.

53. W. K. Clifford, "Body and Mind" (from *Fortnightly Review*). In *Lectures and Essays of the Late William Kingdom Clifford*, ed. L. Stephen and F. Pollock (London: Macmillan, [1874] 1901), 2:38–39.

54. T. H. Huxley, *Lessons in Elementary Physiology* (London: Macmillan, 1866), 210.

55. R. Wright, "Evolution and Higher Purpose," meaningoflife.tv, 2016, http://meaningoflife.tv/articles/wright-evolution-purpose.

56. S. Vogel, *Life's Devices: The Physical World of Animals and Plants* (Princeton, NJ: Princeton University Press 1988).

57. R. Stout, *Action* (Montreal: McGill-Queens University Press, 2005).

58. J. M. Cooper, ed., *Plato: Complete Works* (Indianapolis: Hackett, 1997), 85, 98e–99a.

59. R. Otto, *The Idea of the Holy* (Oxford: Oxford University Press, 1923).

60. P. J. Nahin, *An Imaginary Tale: The Story of* $\sqrt{-1}$ (Princeton, NJ: Princeton University Press, 1998).

61. A. I. Melden, *Free Action* (London: Routledge and Kegan Paul, 1961).

62. G.E.M. Anscombe, *Intention* (Cambridge, MA: Harvard University Press, 2000), 9.

63. Ibid.

64. D. Davidson, "Actions, Reasons and Causes," *Journal of Philosophy* 60 (1963): 3–4.

65. W. Dray, *Laws and Explanation in History* (Oxford: Clarendon, 1957).

66. I. Kershaw, *Hitler, 1936–1945: Nemesis* (New York: Norton, 2000).

67. Davidson, "Actions, Reasons and Causes," 685.

68. Anscombe, *Intention*, 5.

69. Barnes, *The Complete Works of Aristotle*; *Nicomachean Ethics*, 1139a 32–33.

70. I. Kant, *Critique of the Power of Judgment*, trans. and ed. P. Guyer (Cambridge: Cambridge University Press, [1790] 2000), 247.

71. M. Ruse, *The Philosophy of Biology* (London: Hutchinson, 1973).

72. E. Nagel, *The Structure of Science: Problems in the Logic of Scientific Explanation* (New York: Harcourt, Brace and World, 1961).

73. D. Hume, *An Enquiry Concerning Human Understanding*.

74. D. C. Dennett, *Elbow Room: The Varieties of Free Will Worth Wanting* (Cambridge, MA: MIT Press).

75. P. H. Barrett, P. J. Gautrey, S. Herbert, D. Kohn, and S. Smith, eds., *Charles Darwin's Notebooks, 1836–1844* (Ithaca, NY: Cornell University Press, 1987), Notebook M, 27.

76. Ibid.

Chapter Eleven. Religion

1. My copyeditor, Cathy Slovensky, encouraged me to qualify my definition of Christianity, as throughout I focus exclusively on Western forms of Christianity, and, more specifically, on Augustinian Western forms of Christianity. She has a good point, although, I would reply, not a definitive one. The story I am telling is very much one of Western culture—a culture where science has been all-important—and in this context it is Western religion that is equally all-important. But I am sensitive to the challenge, and you

will see that at the end of chapter 12, I show the relevance for me of Quakerism, the non-Augustinian version of Christianity in which I was raised.

2. P. Harvey, *An Introduction to Buddhism: Teachings, History and Practices* (Cambridge: Cambridge University Press, 1990); M. Ruse, *Atheism: What Everyone Needs to Know* (Oxford: Oxford University Press, 2015).

3. Harvey, *An Introduction to Buddhism*, 63.

4. M. Ruse, *The Gaia Hypothesis: Science on a Pagan Planet* (Chicago: University of Chicago Press, 2013).

5. R. Steiner, *Occult Science: An Outline* (Forest Row, Sussex: Rudolf Steiner Press, [1914] 2005).

6. J. Lovelock, *Gaia: A New Look at Life on Earth* (Oxford: Oxford University Press, 1979).

7. L. Lear, *Rachel Carson: Witness for Nature* (New York: Henry Holt, 1997).

8. A. Razak, "Toward a Womanist Analysis of Birth," in *Reweaving the World: The Emergence of Ecofeminism*, ed. I. Diamond and G. F. Orenstein (San Francisco: Sierra Club 1990), 165.

9. P. G. Allen, "The Woman I Love Is a Planet; The Planet I Love Is a Tree," in *Reweaving the World: The Emergence of Ecofeminism*, ed. I. Diamond and G. F. Orenstein (San Francisco: Sierra Club, 1990), 52.

10. O. Zell-Ravenheart, *Green Egg Omelet: An Anthology of Art and Articles from the Legendary Pagan Journal* (Franklin Lakes, NJ: New Page Books, 2009), 92.

11. Ibid., 82.

12. Ibid., 92.

13. J. Edwards, *Sinners in the Hands of an Angry God and Other Puritan Sermons* (New York: Dover, 2005).

14. M. Ruse, *Homosexuality: A Philosophical Inquiry* (Oxford: Blackwell, 1988).

15. Harvey, *An Introduction to Buddhism*, 60–61.

16. R. Carson, *The Edge of the Sea* (Boston: Houghton Mifflin, 1955), 250.

17. Ibid.

18. Razak, "Toward a Womanist Analysis of Birth," 165.

19. Zell-Ravenheart, *Green Egg Omelet*, 93.

20. R. Dawkins, "Religion Is a Virus," *Mother Jones*, 1997; subsequent quotes in this paragraph are from this article.

21. J. Barrett, *Why Would Anyone Believe in God?* (Lanham, MD: AltaMira Press, 2004), 31.

22. A. Plantinga, "Pluralism: A Defense of Religious Exclusivism," in *The*

Philosophical Challenge of Religious Diversity, ed. K. Meeker and P. Quinn (New York: Oxford University Press, 2000).

23. [Saint] Anselm, *Anselm: Proslogium, Monologium; An Appendix on Behalf of the Fool by Gaunilon, and Cur Deus Homo*, trans. S. N. Deane (Chicago: Open Court 1903), 13.

24. T. Aquinas, *Summa Theologica*, vol. 1 (London: Burns, Oates and Washbourne, 1952), 21, 3.

25. J. Polkinghorne, *Belief in God in an Age of Science* (New Haven: Yale University Press, 2003), 106.

26. D. Hume, *An Enquiry Concerning Human Understanding* (Oxford: Oxford University Press, [1748] 2007), 78.

27. C. Darwin, *The Descent of Man, and Selection in Relation to Sex* (London: John Murray, 1871), 1, 67.

28. S. Atran, *In Gods We Trust: The Evolutionary Landscape of Religion* (New York: Oxford University Press, 2004), 78.

29. Ibid.

30. E. Durkheim, *Elementary Forms of Religious Life* (Oxford: Oxford University Press, 1912).

31. E. O. Wilson, *On Human Nature* (Cambridge, MA: Harvard University Press, 1978), 188.

32. Ibid., 178.

33. L. Colley, *Britons: Forging the Nation, 1707–1837* (New Haven: Yale University Press, 1992).

34. M. Ruse, *Atheism*.

35. A. F. Winnington-Ingram, *The Potter and the Clay* (London: Wells, Gardner, and Darton, 1917), 40.

36. P. Appleman, *New and Selected Poems, 1956–1996* (Fayetteville: University of Arkansas Press, 1996).

Chapter Twelve. The End

1. J. S. Mill, *Utilitarianism* (London: Parker, Son, and Bourn, 1863).

2. I. Kant, *Critique of Practical Reason*, trans. T. K. Abbott (London: Longmans, Green, [1788] 1898).

3. M. Ruse, *Philosophy after Darwin* (Princeton, NJ: Princeton University Press, 2009).

4. A. Balfour, *The Foundations of Belief* (New York: Longmans, Green, 1895), 308–9.

5. R. J. Hankinson, *Cause and Explanation in Ancient Greek Thought* (Oxford: Oxford University Press, 1998), 202.

6. A. Plantinga, *Warrant and Proper Function* (New York: Oxford University Press, 1993), 219, quoting from F. Darwin, *The Life and Letters of Charles Darwin, including an Autobiographical Chapter* (London: John Murray, 1887), 1:315–16.

7. Ibid., 223–24.

8. W.V.O. Quine, *Ontological Relativity and Other Essays* (New York: Columbia University Press, 1969), 126.

9. H. Spencer, *Social Statics; or, The Conditions Essential to Human Happiness Specified and the First of Them Developed* (London: J. Chapman, 1851).

10. H. Spencer, *The Data of Ethics* (London: Williams and Norgate, 1879), 21.

11. M. Ruse and E. O. Wilson, "Moral Philosophy as Applied Science," *Philosophy* 61 (1986): 186.

12. E. O. Wilson, *Biophilia* (Cambridge, MA: Harvard University Press, 1984).

13. R. J. Richards, *The Romantic Conception of Life: Science and Philosophy in the Age of Goethe* (Chicago: University of Chicago Press, 1986).

14. M. Ruse, *Taking Darwin Seriously: A Naturalistic Approach to Philosophy* (Oxford: Blackwell, 1986).

15. M. Ruse and E. O. Wilson, "The Evolution of Morality," *New Scientist* 1478 (1985): 855.

16. M. Ruse, *Philosophy after Darwin*; and *Science and Spirituality: Making Room for Faith in the Age of Science* (Cambridge: Cambridge University Press, 2010).

17. Darwin, *The Descent of Man*, 1, 166.

18. Ibid.

19. Ibid.

20. I. Kant, *Foundations of the Metaphysics of Morals*, trans. L. W. Beck (Indianapolis: Bobbs-Merrill, [1785] 1959), 41.

21. Darwin, *The Descent of Man*, 1, 73.

22. P. Singer, "Famine, Affluence and Morality," *Philosophy and Public Affairs* 1 (1972): 229–43.

23. D. Hume, *A Treatise of Human Nature*, ed. D. F. Norton and M. J. Norton (Oxford: Oxford University Press, [1739–40] 2000), 3.2.1.

24. G.W.F. Hegel, *Logic*, trans. W. Wallace (Pacifica, CA: Marxist's Internet Archive, [1830] 2008), 447.

25. J. Mackie, *Ethics* (Harmondsworth: Penguin, 1977).

26. S. Pinker, *The Better Angels of Our Nature: Why Violence Has Declined* (New York: Viking, 2011).

27. J. P. Sartre, *Existentialism Is a Humanism* (New Haven: Yale University Press, 2007), 22.

28. Ibid., 23.

29. S. Wolf, *Meaning in Life and Why It Matters* (Princeton, NJ: Princeton University Press, 2010), 25.

30. Ibid., 3.

31. Sartre, *Existentialism Is a Humanism*, 23.

32. I. Berlin, *The Hedgehog and the Fox* (London: Weidenfeld and Nicolson, 1953).

33. M. Ruse, *But Is It Science? The Philosophical Question in the Creation/Evolution Controversy* (Buffalo, NY: Prometheus, 1988).

34. M. Ruse, *Darwinism as Religion: What Literature Tells Us about Evolution* (Oxford: Oxford University Press, 2017).

35. W. Seager, *Theories of Consciousness: An Introduction and Assessment*, 2nd ed. (London: Routledge 2016), 307.

36. B. Williams, *Moral Luck* (Cambridge: Cambridge University Press, 1981); T. Nagel, *Mortal Questions* (Cambridge: Cambridge University Press, 1979).

Epilogue

1. P. Appleman, "How Evolution Came to Indiana," *Darwin's Ark* (Bloomington: Indiana University Press, [1984] 2009), 65, reprinted with permission of Indiana University Press.

BIBLIOGRAPHY

Africa, T. W. 1961. "Copernicus' Relation to Aristarchus and Pythagoras." *Isis* 52:403–9.

Alexander, S. 1920. *Space, Time and Deity (The Gifford Lectures at Glasgow, 1916–1918).* London: Macmillan.

Allen, C., M. Bekoff, and G. Lauder. 1998. *Nature's Purposes, Analyses of Function and Design in Biology.* Cambridge, MA: MIT Press.

Allen, P. G. 1990. "The Woman I Love Is a Planet; The Planet I Love Is a Tree." In *Reweaving the World: The Emergence of Ecofeminism*, edited by I. Diamond and G. F. Orenstein, 52–57. San Francisco: Sierra Club.

Anscombe, G.E.M. 2000. *Intention.* Cambridge, MA: Harvard University Press.

Anselm [Saint]. 1903. *Anselm: Proslogium, Monologium; An Appendix on Behalf of the Fool by Gaunilon, and Cur Deus Homo.* Translated by S. N. Deane. Chicago: Open Court.

Appleman, P. [1984] 2009. *Darwin's Ark.* Illustrations by R. Pozzatti. Bloomington: Indiana University Press.

———. 1996. *New and Selected Poems, 1956–1996.* Fayetteville: University of Arkansas Press.

Aquinas, [Saint] Thomas. 1952. *Summa Theologica.* Vol. 1. London: Burns, Oates and Washbourne.

———. 1970. *Summa Theologiae.* Vol. 11, *Man (1a. 75–83).* London: Eyre and Spottiswoode.

Arensburg, B., A. M. Tillier, B. Vandermeersch, H. Duday, L. A. Schepartz, and Y. Rak. 1989. "A Middle Paleolithic Human Hyoid Bone." *Nature* 338:758–60.

Aristotle. 1984. *The History of Animals.* Translated by D. W. Thompson. In *The Complete Works of Aristotle*, edited by J. Barnes. Princeton, NJ: Princeton University Press.

Atmanspacher, H. 2015. "Quantum Approaches to Consciousness." *Stanford Encyclopedia of Philosophy.* https://plato.stanford.edu/entries/qt-consciousness/#4.1.

Atran, S. 2004. *In Gods We Trust: The Evolutionary Landscape of Religion.* New York: Oxford University Press.

Augustine. [396] 1998. *Confessions*. Translated by H. Chadwick. Oxford: Oxford University Press.

———. [413–426] 1998. *The City of God against the Pagans*. Edited and translated by R. W. Dyson. Cambridge: Cambridge University Press.

———. 1887. *On the Trinity*. Translated by A. W. Hadden. New York: Philip Schaff.

———. 1955. *On Free Choice of the Will* (*De Libero Arbitrio*). Translated by M. Pontifex. Westminster: Newman Press.

———. 1999. *The Augustine Catechism: The Enchiridion on Faith, Hope and Charity*. Translated by B. Harbert. Hyde Park, NY: New City Press.

———. 2000. *Expositions of the Psalms*. Translated by M. Boulding. Hyde Park, NY: New City Press.

———. 2008. *Responses to Miscellaneous Questions*. Edited by B. Ramsey. Hyde Park, NY: New City Press.

Ayala, F. 1970. "Teleological Explanations in Evolutionary Biology." *Philosophy of Science* 37:1–15.

Bada, J. L., and A. Lazcana. 2009. "The Origin of Life." In *Evolution: The First Four Billion Years*, edited by M. Ruse and J. Travis, 49–79. Cambridge, MA: Harvard University Press.

Baldwin, T. 1990. *G. E. Moore*. London: Routledge and Kegan Paul.

Balfour, A. 1895. *The Foundations of Belief*. New York: Longmans, Green.

Barnes, J., ed. 1984. *The Complete Works of Aristotle*. Princeton, NJ: Princeton University Press.

Barrett, J. 2004. *Why Would Anyone Believe in God?* Lanham, MD: AltaMira Press.

Barrett, P. H., P. J. Gautrey, S. Herbert, D. Kohn, and S. Smith, eds. 1987. *Charles Darwin's Notebooks, 1836–1844*. Ithaca, NY: Cornell University Press.

Barrow, J. D., and F. J. Tipler. 1986. *The Anthropic Cosmological Principle*. Oxford: Clarendon Press.

Barth, K. 1933. *The Epistle to the Romans*. Oxford: Oxford University Press.

Bedau, M. 1992. "Where's the Good in Teleology?" *Philosophy and Phenomenological Research* 52:781–805.

Behe, M. 1996. *Darwin's Black Box: The Biochemical Challenge to Evolution*. New York: Free Press.

Bergson, H. 1907. *L'évolution créatrice*. Paris: Alcan.

———. 1911. *Creative Evolution*. New York: Holt.

Berlin, I. 1953. *The Hedgehog and the Fox*. London: Weidenfeld and Nicolson.

Bowler, P. J. 1984. *Evolution: The History of an Idea*. Berkeley: University of California Press.

Boyle, R. [1688] 1966. "A Disquisition about the Final Causes of Natural Things." In *The Works of Robert Boyle*, edited by T. Birch. Vol. 5. Hildesheim: Georg Olms.

———. 1996. *A Free Enquiry into the Vulgarly Received Notion of Nature*. Edited by E. B. Davis and M. Hunter. Cambridge: Cambridge University Press.

Browne, J. 1995. *Charles Darwin: Voyaging. Volume 1 of a Biography*. New York: Knopf.

———. 2002. *Charles Darwin: The Power of Place. Volume 2 of a Biography*. New York: Knopf.

Brown, P. 1967. *Augustine of Hippo: A Biography*. London: Faber and Faber.

Burkhardt, R. W. 1977. *The Spirit of System: Lamarck and Evolutionary Biology*. Cambridge, MA: Harvard University Press.

Burtt, E. A. 1932. *The Metaphysical Foundations of Modern Science*. New York: Harcourt, Brace.

Bury, J. B. [1920] 1924. *The Idea of Progress: An Inquiry into Its Origin and Growth*. London: Macmillan.

Carroll, S. B. 2005. *Endless Forms Most Beautiful: The New Science of Evo Devo*. New York: Norton.

Carson, R. 1955. *The Edge of the Sea*. Boston: Houghton Mifflin.

———. 1962. *Silent Spring*. New York: Houghton Mifflin.

Chalmers, D. J. 1996. *The Conscious Mind*. New York: Oxford University Press.

———. 1997. "Facing Up to the Problem of Consciousness." In *Explaining Consciousness: The "Hard Problem."* Edited by J. Shear, 9–32. Cambridge, MA: MIT Press.

Chambers, R. 1844. *Vestiges of the Natural History of Creation*. London: J. Churchill.

———. 1846. *Vestiges of the Natural History of Creation*. 5th ed. London: J. Churchill.

Chomsky, N. 1957. *Syntactic Structures*. The Hague: Mouton.

Cicero. 1933. *De Natura Deorum*. Translated by H. Rackham. Cambridge, MA: Harvard University Press, Loeb Classical Library.

Clifford, W. K. 1901. "Body and Mind" (from *Fortnightly Review*). In *Lectures and Essays of the Late William Kingdom Clifford*, edited by L. Stephen and F. Pollock, 1–51. Vol. 2. London: Macmillan.

Coleman, W. 1964. *Georges Cuvier Zoologist: A Study in the History of Evolution Theory*. Cambridge, MA: Harvard University Press.

Colley, L. 1992. *Britons: Forging the Nation, 1707–1837*. New Haven: Yale University Press.

Conroy, G. C., and H. Pontzer. 2012. *Reconstructing Human Origins: A Modern Synthesis*. 3rd ed. New York: Norton.

Conway Morris, S. 2003. *Life's Solution: Inevitable Humans in a Lonely Universe*. Cambridge: Cambridge University Press.

Cooper, J. M., ed. 1997. *Plato: Complete Works*. Indianapolis: Hackett.

Coyne, J. A. 2009. *Why Evolution is True*. New York: Viking.

Coyne, J. A., and H. A. Orr. 2004. *Speciation*. Sunderland, MA: Sinauer.

Coyne, J. A., N. H. Barton, and M. Turelli. 1997. "Perspective: A Critique of Sewall Wright's Shifting Balance Theory of Evolution." *Evolution* 51, no. 3:643–71.

Cummins, R. 1975. "Functional Analysis." *Journal of Philosophy* 72, no. 20 (Nov. 20): 741–65.

Cunningham, S. 1996. *Philosophy and the Darwinian Legacy*. Rochester: University of Rochester Press.

Cuvier, G. 1817. *Le règne animal distribué d'aprés son organisation, pour servir de base à l'histoire naturelle des animaux et d'introduction à l'anatomie comparée*. Paris.

———. 1831. *The Animal Kingdom (Le règne animal distribué d'aprés son organisation, pour servir de base à l'histoire naturelle des animaux et d'introduction à l'anatomie comparée)*. Translated by H. M'Murtrie. New York: Carvill.

Darwin, C. 1839. *Journal of Researches into the Geology and Natural History of the Various Countries Visited by HMS Beagle*. London: Henry Colburn.

———. 1859. *On the Origin of Species by Means of Natural Selection, or the Preservation of Favoured Races in the Struggle for Life*. London: John Murray.

———. 1861. *Origin of Species*. 3rd ed. London: John Murray.

———. 1862. *On the Various Contrivances by which British and Foreign Orchids are Fertilized by Insects, and On the Good Effects of Intercrossing*. London: John Murray.

———. 1868. *The Variation of Animals and Plants under Domestication*. London: Murray.

———. 1871. *The Descent of Man, and Selection in Relation to Sex*. London: John Murray.

———. 1958. *The Autobiography of Charles Darwin, 1809–1882*. Edited by Nora Barlow. London: Collins.

———. 1985–. *The Correspondence of Charles Darwin*. Cambridge: Cambridge University Press.

Darwin, C., and A. R. Wallace. 1958. *Evolution by Natural Selection.* Foreword by Gavin de Beer. Cambridge: Cambridge University Press.

Darwin, E. [1794–96] 1801. *Zoonomia; or, The Laws of Organic Life.* 3rd ed. London: J. Johnson.

———. 1803. *The Temple of Nature.* London: J. Johnson.

Darwin, F. 1887. *The Life and Letters of Charles Darwin, including an Autobiographical Chapter.* London: John Murray.

Davidson, D. 1963. "Actions, Reasons and Causes." *Journal of Philosophy* 60:685–700.

Dawkins, R. 1976. *The Selfish Gene.* Oxford: Oxford University Press.

———. 1982. *The Extended Phenotype: The Gene as the Unit of Selection.* Oxford: W. H. Freeman.

———. 1986. *The Blind Watchmaker.* New York: Norton.

———. 1992. "Progress." In *Keywords in Evolutionary Biology.* Edited by E. F. Keller and E. Lloyd, 263–72. Cambridge, MA: Harvard University Press.

———. 1997. "Religion Is a Virus." *Mother Jones.*

Dawkins, R., and J. R. Krebs. 1979. "Arms Races between and within Species." *Proceedings of the Royal Society of London, B* 205:489–511.

Dennett, D. C. 1984. *Elbow Room: The Varieties of Free Will Worth Wanting.* Cambridge, MA: MIT Press.

Descartes, R. [1642] 1964. "Meditations." In *Philosophical Essays,* 59–143. Indianapolis: Bobbs-Merrill.

Desmond, A. 1997. *Huxley: From Devil's Disciple to Evolution's High Priest.* New York: Basic Books.

Dibben, M. 2004. "Exploring the Processual Nature of Trust and Cooperation in Organisations: A Whiteheadian Analysis." *Philosophy of Management* 4:25–39.

Diderot, D. [1796] 1972. *The Nun.* London: Penguin.

———. 1943. *Diderot: Interpreter of Nature.* New York: International.

Dijksterhuis, E. J. 1961. *The Mechanization of the World Picture.* Oxford: Oxford University Press.

Dixon, M., and G. Radick. 2009. *Darwin in Ilkley.* Stroud, Gloucestershire: History Press.

Dobzhansky, T., F. J. Ayala, G. L. Stebbins, and J. W. Valentine. 1977. *Evolution.* San Francisco: Freeman.

Doolittle, R. F. 1997. "A Delicate Balance." *Boston Review* 22, no. 1:28–29.

Doolittle, W. F. 1981. "Is Nature Really Motherly?" *CoEvolution* 29:58–62.

Doyle, J. A. 1977. "Patterns of Evolution in Early Angiosperms." In *Patterns of Evolution as Illustrated by the Fossil Record,* edited by A. Hallam, 501–42. Amsterdam: Elsevier.

Dray, W. 1957. *Laws and Explanation in History*. Oxford: Clarendon.

Driesch, H. 1908. *The Science and the Philosophy of the Organism*. London: Adam and Charles Black.

Durkheim, E. 1912. *Elementary Forms of Religious Life*. Oxford: Oxford University Press.

Edwards, J. 2005. *Sinners in the Hands of an Angry God and Other Puritan Sermons*. New York: Dover.

El-Sabaawi, R. W., M. C. Marshall, R. D. Bassar, A. López-Sepulcre, E. P. Palkovacs, and C. Dalton. 2015. "Assessing the Effects of Guppy Life History Evolution on Nutrient Recycling: From Experiments to the Field." *Freshwater Biology* 60:590–601.

Falk, D. 2004. *Braindance: New Discoveries about Human Origins and Brain Evolution*. Gainesville: University of Florida Press.

———. 2012. *The Fossil Chronicles: How Two Controversial Discoveries Changed Our View of Human Evolution*. Berkeley: University of California Press.

Farlow, J. O., C. V. Thompson, and D. E. Rosner. 1976. "Plates of the Dinosaur Stegosaurus: Forced Convection Heat Loss Fins?" *Science* 192: 1123–25.

Fodor, J., and M. Piattelli-Palmarini. 2010. *What Darwin Got Wrong*. New York: Farrar, Straus and Giroux.

Fortey, R. 2005. *The Earth: An Intimate History*. New York: Vintage.

Galileo. [1632] 2001. *Dialogue Concerning the Two Chief World Systems*. Translated by S. Drake. New York: Random House.

Gillespie, C. C. 1950. *Genesis and Geology*. Cambridge, MA: Harvard University Press.

Godwin, W. [1793] 2013. *An Enquiry Concerning Political Justice*. Oxford: Oxford University Press.

Golding, W. [1955] 1962. *The Inheritors*. New York: Harcourt.

Goodwin, B. 2001. *How the Leopard Changed Its Spots*. 2nd ed. Princeton, NJ: Princeton University Press.

Gotthelf, A. 1999. "Darwin on Aristotle." *Journal of the History of Biology* 32:3–30.

Gould, S. J. 1971. "D'Arcy Thompson and the Science of Form." *New Literary History* 2:229–58.

———. 1977. *Ontogeny and Phylogeny*. Cambridge, MA: Belknap Press.

———. 1980. "The Piltdown Conspiracy." *Natural History* 89 (August): 8–28.

———. 1981. *The Mismeasure of Man*. New York: Norton.

———. 1985. *The Flamingo's Smile: Reflections in Natural History*. New York: Norton.

———. 1988. "On Replacing the Idea of Progress with an Operational Notion of Directionality." In *Evolutionary Progress*, edited by M. H. Nitecki, 319–38. Chicago: University of Chicago Press.

———. 1989. *Wonderful Life: The Burgess Shale and the Nature of History*. New York: W. W. Norton.

———. 1996. *Full House: The Spread of Excellence from Plato to Darwin*. New York: Paragon.

———. 2002. *The Structure of Evolutionary Theory*. Cambridge, MA: Harvard University Press.

Gould, S. J., and R. C. Lewontin. 1979. "The Spandrels of San Marco and the Panglossian Paradigm: A Critique of the Adaptationist Programme." *Proceedings of the Royal Society of London, Series B: Biological Sciences* 205:581–98.

Grant, P. R. 1986. *Ecology and Evolution of Darwin's Finches*. Princeton, NJ: Princeton University Press.

Grant, P. R., and R. B. Grant. 2007. *How and Why Species Multiply: The Radiation of Darwin's Finches*. Princeton, NJ: Princeton University Press.

Gray, A. [1860] 1876. Review of *The Origin of Species by Means of Natural Selection, American Journal of Arts and Sciences*. In *Darwiniana*, 7–50. New York: Appleton.

———. 1881. *Structural Botany*. 6th ed. London: Macmillan.

Haeckel, E. 1892. "Our Monism: The Principles of a Consistent, Unitary World-View." *Monist* 2:481–86.

Haldane, J.B.S. 1927. *Possible Worlds and Other Essays*. London: Chatto and Windus.

Hall, A. R. 1954. *The Scientific Revolution, 1500–1800: The Formation of the Modern Scientific Attitude*. London: Longman, Green.

———. 1990. *Henry More: Magic, Religion and Experiment*. Oxford: Blackwell.

Hankinson, R. J. 1998. *Cause and Explanation in Ancient Greek Thought*. Oxford: Oxford University Press.

Harvey, P. 1990. *An Introduction to Buddhism: Teachings, History and Practices*. Cambridge: Cambridge University Press.

Hegel, G.W.F. [1817] 1970. *Philosophy of Nature*. Oxford: Oxford University Press.

———. [1830] 2008. *Logic*. Translated by W. Wallace. Pacifica, CA: Marxist's Internet Archive.

Henry, W. 1818. *The Elements of Experimental Chemistry*. 8th ed. London: Baldwin, Cradock and Joy.

Holden, C. 2004. "The Origin of Speech." *Science* 303:1316–19.

Holton, G. 1973. *Thematic Origins of Scientific Thought*. Cambridge, MA: Harvard University Press.

Hopson, J. A. 1975. "The Evolution of Cranial Display Structures in Hadrosaurian Dinosaurs." *Paleobiology* 1:21–43.

Hume, D. [1739–40] 2000. *A Treatise of Human Nature*. Edited by D. F. Norton and M. J. Norton. Oxford: Oxford University Press.

———. [1748] 2007. *An Enquiry Concerning Human Understanding*. Oxford: Oxford University Press.

———. [1757] 1963. "A Natural History of Religion." In *Hume on Religion*, edited by R. Wollheim. London: Fontana.

———. [1779] 1990. *Dialogues Concerning Natural Religion*. Edited by M. Bell. London: Penguin.

Hursthouse, R. 1999. *On Virtue Ethics*. Oxford: Oxford University Press.

Huxley, J. S. 1912. *The Individual in the Animal Kingdom*. Cambridge: Cambridge University Press.

———. 1927. *Religion without Revelation*. London: Ernest Benn.

———. 1942. *Evolution: The Modern Synthesis*. London: Allen and Unwin.

Huxley, T. H. 1866. *Lessons in Elementary Physiology*. London: Macmillan.

———. 1874. "On the Hypothesis that Animals Are Automata, and Its History." *Fortnightly Review* 16:555–80.

Jacob, F. 1977. "Evolution and Tinkering." *Science* 196:1161–66.

James, W. 1880a. "Great Men, Great Thoughts, and the Environment." *Atlantic Monthly* 46 (276): 441–59.

———. 1880b. *The Principles of Psychology*. New York: Henry Holt.

Jantzen, B. C. 2014. *An Introduction to Design Arguments*. Cambridge: Cambridge University Press.

Jerison, H. 1973. *Evolution of the Brain and Intelligence*. New York: Academic Press.

Johanson, D., and M. Edey. 1981. *Lucy: The Beginnings of Humankind*. New York: Simon and Schuster.

John Paul II. 1997. "The Pope's Message on Evolution." *Quarterly Review of Biology* 72:377–83.

Johnson, M. R. 2005. *Aristotle on Teleology*. Oxford: Oxford University Press.

Johnson, P. E. 1991. *Darwin on Trial*. Washington, DC: Regnery Gateway.

Kant, I. [1781/1787] 1998. *Critique of Pure Reason*. Translated and edited by P. Guyer and A. W. Wood. Cambridge: Cambridge University Press.

———. [1785] 1959. *Foundations of the Metaphysics of Morals*. Translated by L. W. Beck. Indianapolis: Bobbs-Merrill.

———. [1786] 2004. *Metaphysical Foundations of Natural Science*. Translated by M. Friedman. Cambridge: Cambridge University Press.

———. [1788] 1898. *Critique of Practical Reason*. Translated by T. K. Abbott. London: Longmans, Green.

———. [1790] 2000. *Critique of the Power of Judgment*. Translated and edited by P. Guyer. Cambridge: Cambridge University Press.

———. [1793] 1998. *Religion within the Bounds of Mere Reason*. Edited by A. Wood and G. di Giovanni. Cambridge: Cambridge University Press.

Kauffman, S. A. 1993. *The Origins of Order: Self-Organization and Selection in Evolution*. Oxford: Oxford University Press.

———. 1995. *At Home in the Universe: The Search for the Laws of Self-Organization and Complexity*. New York: Oxford University Press.

Kepler, J. 1977. *The Harmony of the World*. Translated by E. J. Aiton, A. M. Duncan, and J. V. Field. Philadelphia: American Philosophical Society.

Kershaw, I. 2000. *Hitler, 1936–1945: Nemesis*. New York: Norton.

Kierkegaard, S. [1843] 1985. *Fear and Trembling*. Edited and translated by A. Hannay. London: Penguin.

Kim, J. 1993. *Supervenience and Mind: Selected Philosophical Essays*. Cambridge: Cambridge University Press.

Kimura, M. 1983. *The Neutral Theory of Molecular Evolution*. Cambridge: Cambridge University Press.

King, D. 1996. "An Interview with Professor Brian Goodwin." *GenEthics News* 11:6–8.

Kirby, W., and W. Spence. 1815–28. *An Introduction to Entomology: or, Elements of the Natural History of Insects*. London: Longman, Hurst, Reece, Orme, and Brown.

Kirk, G. M., J. E. Raven, and M. Schofield. 1984. *The Presocratic Philosophers: A Critical History with a Selection of Texts*. Cambridge: Cambridge University Press.

Krings, M., A. Stone, R. W. Schmitz, H. Krainitzki, M. Stoneking, and S. Pääbo. 1997. "Neanderthal DNA Sequences and the Origin of Modern Humans." *Cell* 90:19–30.

Kuhn, T. 1962. *The Structure of Scientific Revolutions*. Chicago: University of Chicago Press.

Lamarck, J.-B. 1809. *Philosophie zoologique*. Paris: Dentu.

Lawrence, D. H. [1915] 1949. *The Rainbow*. London: Penguin.

———. [1921] 1960. *Women in Love*. London: Penguin.

Lear, L. 1997. *Rachel Carson: Witness for Nature.* New York: Henry Holt.

Leibniz, G.F.W. 1714. *Monadology and Other Philosophical Essays.* New York: Bobbs-Merrill.

Lennox, J. G. 2001. *Aristotle's Philosophy of Biology.* Cambridge: Cambridge University Press.

Leunissen, M. 2010. *Explanation and Teleology in Aristotle's Science of Nature.* Cambridge: Cambridge University Press.

Lewontin, R. C. 1974. *The Genetic Basis of Evolutionary Change.* New York: Columbia University Press.

Lieberman, P. 1984. *The Biology and Evolution of Language.* Cambridge, MA: Harvard University Press.

Lindberg, D. C. 1992. *The Beginnings of Western Science: The European Scientific Tradition in Philosophical, Religious, and Institutional Context, Prehistory to A.D. 1450.* Chicago: University of Chicago Press.

Lipton, P. 1991. *Inference to the Best Explanation.* London: Routledge.

Lockwood, M. 1989. *Mind, Brain and the Quantum: The Compound "I."* Oxford: Blackwell.

Lovelock, J. E. 1979. *Gaia: A New Look at Life on Earth.* Oxford: Oxford University Press.

Lucretius. 1950. *Of the Nature of Things.* Translated by W. E. Leonard. London: Dutton, Everyman's Library.

Lurie, E. 1960. *Louis Agassiz: A Life in Science.* Chicago: University of Chicago Press.

Mackie, J. 1977. *Ethics.* Harmondsworth: Penguin.

Malthus, T. R. [1798] 1966. *An Essay on the Principle of Population.* New York: Macmillan.

Maynard Smith, J. 1981. "Did Darwin Get It Right?" *London Review of Books* 3, 11:10–11.

———. 1995. "Genes, Memes, and Minds." *New York Review of Books* 42, no. 19:46–48.

McShea, D., and R. Brandon. 2010. *Biology's First Law: The Tendency for Diversity and Complexity to Increase in Evolutionary Systems.* Chicago: University of Chicago Press.

Medawar, P. B. 1961. Review of *The Phenomenon of Man. Mind* 70:99–106.

Melden, A. I. 1961. *Free Action.* London: Routledge and Kegan Paul.

Mill, J. S. 1863. *Utilitarianism.* London: Parker, Son, and Bourn.

Miller, K. 1999. *Finding Darwin's God.* New York: Harper and Row.

Mithen, S. 1996. *The Prehistory of the Mind.* London: Thames and Hudson.

Moore, G. E. 1903. *Principia Ethica.* Cambridge: Cambridge University Press.

More, H. [1659] 1987. *The Immortality of the Soul.* Dordrecht: Nijhoff.

Morris, H. M. 1999. "Design Is Not Enough." *Back to Genesis* 127:a–c.

Morwood, M., and P. Van Oosterzee. 2007. *A New Human: The Startling Discovery and Strange Story of the "Hobbits" of Flores, Indonesia.* London: Collins.

Mossio, M., C. Saborido, A. Moreno. 2009. "An Organizational Account of Biological Functions." *British Journal for the Philosophy of Science* 60, no. 4:813–41.

Naden, C. 1999. *Poetical Works of Constance Naden.* Kernville, CA: High Sierra Books.

Nagel, E. 1961. *The Structure of Science: Problems in the Logic of Scientific Explanation.* New York: Harcourt, Brace and World.

Nagel, T. 1979. *Mortal Questions.* Cambridge: Cambridge University Press.

———. 2012. *Mind and Cosmos: Why the Materialist Neo-Darwinian Conception of Nature Is Almost Certainly False.* New York: Oxford University Press.

Nahin, P. J. 1998. *An Imaginary Tale: The Story of* $\sqrt{-1}$. Princeton, NJ: Princeton University Press.

Noll, M. 2002. *America's God: From Jonathan Edwards to Abraham Lincoln.* New York: Oxford University Press.

Noonan, J. P., G. Coop, S. Kudaravalli, D. Smith, J. Krause, J. Alessi, F. Chen, D. Platt, S. Pääbo, J. Pritchard, and E. Rubin. 2006. "Sequencing and Analysis of Neanderthal Genomic DNA." *Science* 314:1113–18.

Numbers, R. L. 2006. *The Creationists: From Scientific Creationism to Intelligent Design.* Cambridge, MA: Harvard University Press.

Otto, R. 1923. *The Idea of the Holy.* Oxford: Oxford University Press.

Outram, D. 1984. *Georges Cuvier: Vocation, Science and Authority in Post-Revolutionary France.* Manchester: Manchester University Press.

Owen, R. 1848. *On the Archetype and Homologies of the Vertebrate Skeleton.* London: Voorst.

———. 1849. *On the Nature of Limbs.* London: Voorst.

Paley, W. [1802] 1819. *Natural Theology* (*Collected Works: Volume 4*). London: Rivington.

Parkes, S. 1818. *The Chemical Catechism, with Notes, Illustrations and Experiments.* London: Baldwin, Cradock and Joy.

Pedersen, O. 1993. *Early Physics and Astronomy: A Historical Introduction.* Cambridge: Cambridge University Press.

Pinker, S. 1994. *The Language Instinct: How the Mind Creates Language.* New York: William Morrow.

———. 1997. *How the Mind Works.* New York: Norton.

Pinker, S. 2011. *The Better Angels of Our Nature: Why Violence Has Declined.* New York: Viking.

Plantinga, A. 1991. "Where Faith and Reason Clash: Evolution and the Bible," *Christian Scholars Review*, 21, 8–32.

———. 1993. *Warrant and Proper Function.* New York: Oxford University Press.

———. 2000. "Pluralism: A Defense of Religious Exclusivism." In *The Philosophical Challenge of Religious Diversity*, edited by K. Meeker and P. Quinn, 172–92. New York: Oxford University Press.

———. 2004. "Supralapsarianism, or 'O Felix Culpa.' " In *Christian Faith and the Problem of Evil*, edited by P. Van Inwagen, 1–25. Grand Rapids, MI: Eerdmans.

———. 2011. *Where the Conflict Really Lies: Science, Religion, and Naturalism.* New York: Oxford University Press.

Plotinus. 1992. *The Enneads.* Translated by S. MacKenna. Burdett, NY: Larson Publications.

Polkinghorne, J. 2003. *Belief in God in an Age of Science.* New Haven: Yale University Press.

Pollan, M. 2013. "The Intelligent Plant: Scientists Debate a New Way of Understanding Flora." *New Yorker* (December): 92–105.

Provine, W. B. 1971. *The Origins of Theoretical Population Genetics.* Chicago: University of Chicago Press.

———. 1986. *Sewall Wright and Evolutionary Biology.* Chicago: University of Chicago Press.

Quine, W.V.O. 1969. *Ontological Relativity and Other Essays.* New York: Columbia University Press.

Rawls, J. 1971. *A Theory of Justice.* Cambridge, MA: Harvard University Press.

Ray, J. 1691. *Wisdom of God, Manifested in the Words of Creation.* London: William Innys. The standard edition (7th) was published in 1717.

Razak, A. 1990. "Toward a Womanist Analysis of Birth." In *Reweaving the World: The Emergence of Ecofeminism*, edited by I. Diamond and G. F. Orenstein, 165–72. San Francisco: Sierra Club.

Reynolds, V., and R. Tanner. 1983. *The Biology of Religion.* London: Longman.

Reznick, D. N. 2009. *The "Origin" Then and Now: An Interpretive Guide to the "Origin of Species."* Princeton, NJ: Princeton University Press.

———. 2011. "Guppies and the Empirical Study of Adaptation." In *In Light of Evolution: Essays from the Laboratory and Field*, edited by J. B. Losos, 205–32. Greenwood Village, CO: Roberts.

Rhees, R., ed. 1981. *Ludwig Wittgenstein: Personal Recollections*. Oxford: Blackwell.

Richards, R. J. 2002. *The Romantic Conception of Life: Science and Philosophy in the Age of Goethe*. Chicago: University of Chicago Press.

Richards, R. J., and M. Ruse. 2016. *Debating Darwin*. Chicago: University of Chicago Press.

Rogers, P. 2001. *Song of the World Becoming: New and Collected Poems, 1981–2001*. Minneapolis: Milkweed.

Ruse, M. 1973. *The Philosophy of Biology*. London: Hutchinson.

———. 1975. "Darwin's Debt to Philosophy: An Examination of the Influence of the Philosophical Ideas of John F. W. Herschel and William Whewell on the Development of Charles Darwin's Theory of Evolution." *Studies in History and Philosophy of Science* 6:159–81.

———. 1979. *The Darwinian Revolution: Science Red in Tooth and Claw*. Chicago: University of Chicago Press.

———. 1982. *Darwinism Defended: A Guide to the Evolution Controversies*. Reading, MA: Benjamin/Cummings.

———. 1986. *Taking Darwin Seriously: A Naturalistic Approach to Philosophy*. Oxford: Blackwell.

———, ed. 1988a. *But Is It Science? The Philosophical Question in the Creation/Evolution Controversy*. Buffalo, NY: Prometheus.

———. 1988b. *Homosexuality: A Philosophical Inquiry*. Oxford: Blackwell.

———. 1996. *Monad to Man: The Concept of Progress in Evolutionary Biology*. Cambridge, MA: Harvard University Press.

———. 2003. *Darwin and Design: Does Evolution Have a Purpose?* Cambridge, MA: Harvard University Press.

———. 2005. *The Evolution-Creation Struggle*. Cambridge, MA: Harvard University Press.

———. 2006. "Kant and Evolution." In *Theories of Generation*, edited by J. Smith, 402–15. Cambridge: University of Cambridge Press.

———. 2009. *Philosophy after Darwin*. Princeton, NJ: Princeton University Press.

———. 2010. *Science and Spirituality: Making Room for Faith in the Age of Science*. Cambridge: Cambridge University Press.

———. 2011. "The Shame of Calvin College." *Brainstorm: Chronicle of Higher Education*. http://www.chronicle.com/blogs/brainstorm/the-shame-of-calvin-college/37484.

———. 2012. *The Philosophy of Human Evolution*. Cambridge: Cambridge University Press.

Ruse, M.. 2013. *The Gaia Hypothesis: Science on a Pagan Planet.* Chicago: University of Chicago Press.

———. 2015a. "Sexual Selection: Why Does It Play Such a Large Role in the *Descent of Man*?" In *Current Perspectives on Sexual Selection: What's Left after Darwin?*, edited by T. Hoquet, 3–17. New York: Springer.

———. 2015b. *Atheism: What Everyone Needs to Know.* Oxford: Oxford University Press.

———. 2017. *Darwinism as Religion: What Literature Tells Us about Evolution.* Oxford: Oxford University Press.

Ruse, M., and E. O. Wilson. 1985. "The Evolution of Morality." *New Scientist* 1478:108–28.

———. 1986. "Moral Philosophy as Applied Science." *Philosophy* 61:173–92.

Russell, B. 1914. *Our Knowledge of the External World as a Field for Scientific Method in Philosophy.* Chicago: Open Court.

———. 1938. *Power: A New Social Analysis.* London: Allen and Unwin.

———. 1959. *My Philosophical Development.* London: Allen and Unwin.

Russell, R. J. 2008. *Cosmology: From Alpha to Omega: The Creative Mutual Interaction of Theology and Science.* Minneapolis: Fortress Press.

Sartre, J. P. 2007. *Existentialism Is a Humanism.* New Haven: Yale University Press.

Schelling, F.W.J. [von]. [1803] 1988. *Ideas for a Philosophy of Nature as Introduction to the Study of This Science, 1797.* Translated by E. E. Harris and P. Heath. 2nd ed. Cambridge: Cambridge University Press.

Seager, W. 2016. *Theories of Consciousness: An Introduction and Assessment.* 2nd ed. London: Routledge.

Sedgwick, A. 1831. "Address to the Geological Society." *Proceedings of the Geological Society of London* 1:281–316.

Sedley, D. 2007. *Creationism and Its Critics in Antiquity.* Berkeley: University of California Press.

Shipman, P. 2002. *The Man Who Found the Missing Link: Eugene Dubois and His Lifelong Quest to Prove Darwin Right.* Cambridge, MA: Harvard University Press.

Sidgwick, H. 1874. *The Methods of Ethics.* London: Macmillan.

———. 1876. "The Theory of Evolution in Its Application to Practice." *Mind* 1:52–67.

Singer, P. 1972. "Famine, Affluence and Morality." *Philosophy and Public Affairs* 1:229–43.

Skrbina, D. 2005. *Panpsychism in the West.* Cambridge, MA: MIT Press.

Smart, J.J.C. 1959. "Sensations and Brain Processes." *Philosophical Review* 68:141–56.

Smith, A. M. 1987. "Descartes's Theory of Light and Refraction: A Discourse on Method." *Transactions of the American Philosophical Society* 77:1–92.

Sober, E. 2014. "Evolutionary Theory, Causal Completeness, and Theism: The Case of 'Guided' Mutations." In *Evolutionary Biology: Conceptual, Ethical, and Religious Issues*, edited by R. P. Thompson and D. M. Walsh, 31–44. Cambridge: Cambridge University Press.

Spencer, H. 1851. *Social Statics; or, The Conditions Essential to Human Happiness Specified and the First of Them Developed*. London: J. Chapman.

———. 1852. "A Theory of Population, Deduced from the General Law of Animal Fertility." *Westminster Review* 1:468–501.

———. 1857. "Progress: Its Law and Cause." *Westminster Review* 67: 244–67.

———. 1860. "The Social Organism." *Westminster Review*.

———. 1862. *First Principles*. London: Williams and Norgate.

———. 1879. *The Data of Ethics*. London: Williams and Norgate.

———. 1904. *Autobiography*. London: Williams and Norgate.

Spinoza, B. 1985. "Ethics." In *The Collected Writings of Spinoza*, translated by E. Curley. Princeton, NJ: Princeton University Press.

Steiner, R. [1914] 2005. *Occult Science: An Outline*. Forest Row, Sussex: Rudolf Steiner Press.

Stenger, V. J. 2011. *The Fallacy of Fine-Tuning: Why the Universe Is Not Designed for Us*. Buffalo, NY: Prometheus.

Stout, R. 2005. *Action*. Montreal: McGill-Queens University Press.

Strawson, G., et al. 2006. *Consciousness and Its Place in Nature: Does Physicalism Entail Panpsychism?* Exeter: Imprint Academic.

Stringer, C. 2002. "Modern Human Origins: Progress and Prospects." *Philosophical Transactions of the Royal Society, London (B)* 357:563–79.

———. 2003. "Human Evolution: Out of Ethiopia." *Nature* 423:692–95.

Teilhard de Chardin, P. 1955. *Le phénomène humain*. Paris: Editions de Seuil.

———. [1955] 1959. *The Phenomenon of Man*. London: Collins.

Tennyson, A. [1850] 1973. "In Memoriam." In *In Memoriam: An Authoritative Text, Backgrounds and Sources, Criticism*, edited by R. H. Ross, 3–90. New York: Norton.

Thompson, D. W. 1948. *On Growth and Form*. 2nd ed. Cambridge: Cambridge University Press.

Thomson, K. S. "The Pattern of Diversification among Fishes." In *Patterns of Evolution as Illustrated by the Fossil Record*, edited by A. Hallam, 5:547–62. Amsterdam: Elsevier.

Vogel, S. 1988. *Life's Devices: The Physical World of Animals and Plants.* Princeton, NJ: Princeton University Press.

Voltaire. [1759] 1950. *Candide.* Harmondsworth: Penguin.

Vonnegut, K. [1985] 1999. *Galapagos.* New York: Random House.

Wallace, A. R. [1865] 1870. "The Limits of Natural Selection as Applied to Man." In *Contributions to the Theory of Natural Selection.* London: Macmillan.

Weinberg, Steven. 1999. "A Designer Universe?" *New York Review of Books,* October 21.

Weishampel, D. B. 1981. "Acoustic Analyses of Potential Vocalization in Lambeosaurine Dinosaurs (Reptilia: Ornithischia)." *Paleobiology* 7:252–61.

———. 1997. "Dinosaurian Cacophony." *BioScience* 47, no. 3:150–58.

Whewell, W. 1833. *Astronomy and General Physics* (*Bridgewater Treatise, 3*). London: William Pickering.

———. 1837. *The History of the Inductive Sciences.* London: Parker.

———. 1840. *The Philosophy of the Inductive Sciences.* London: Parker.

———. 2001. *Of the Plurality of Worlds: A Facsimile of the First Edition of 1853, Plus Previously Unpublished Material Excised by the Author Just Before the Book Went to Press; and Whewell's Dialogue Rebutting His Critics, Reprinted from the Second Edition.* Edited by M. Ruse. Chicago: University of Chicago Press.

Whitcomb, J. C., and H. M. Morris. 1961. *The Genesis Flood: The Biblical Record and Its Scientific Implications.* Philadelphia: Presbyterian and Reformed Publishing Company.

Whitehead, A. N. [1925] 1967. *Science and the Modern World.* New York: Free Press.

———. [1929] 1978. *Process and Reality: An Essay in Cosmology.* New York: Free Press.

———. 1933. *Adventures of Ideas.* New York: Macmillan.

White, T. D., B. Asfaw, Y. Beyene, Y. Haile-Selassie, C. O. Lovejoy, G. Suwa, and G. Woldegabriel. 2009. "*Ardipithecus ramidus* and the Paleobiology of Early Hominids." *Science* 326 (5949): 75–86.

Williams, B. 1973. "The Makropulos Case: Reflections on the Tedium of Immortality." In *Problems of the Self.* Cambridge: Cambridge University Press.

———. 1981. *Moral Luck.* Cambridge: Cambridge University Press.

———. 1985. *Ethics and the Limits of Philosophy.* London: Fontana.

Williams, G. C. 1966. *Adaptation and Natural Selection.* Princeton, NJ: Princeton University Press.

Wilson, E. O. 1975. *Sociobiology: The New Synthesis*. Cambridge, MA: Harvard University Press.

———. 1978. *On Human Nature*. Cambridge, MA: Harvard University Press.

———. 1984. *Biophilia*. Cambridge, MA: Harvard University Press.

———. 1992. *The Diversity of Life*. Cambridge, MA: Harvard University Press.

Winnington-Ingram, A. F. 1917. *The Potter and the Clay*. London: Wells, Gardner, and Darton.

Wittgenstein, L. 1922. *Tractatus Logico-Philosophicus*. London: Routledge & Kegan Paul.

Wolf, S. 2010. *Meaning in Life and Why It Matters*. Princeton, NJ: Princeton University Press.

Wright, L. "Functions." *Philosophical Review* 82, no. 2 (April 1973): 139–68.

Wright, R. 2016. "Evolution and Higher Purpose." meaningoflife.tv. http://meaningoflife.tv/articles/wright-evolution-purpose.

Wright, S. 1931. "Evolution in Mendelian Populations." *Genetics* 16:97–159.

———. 1932. "The Roles of Mutation, Inbreeding, Crossbreeding and Selection in Evolution." *Proceedings of the Sixth International Congress of Genetics* 1:356–66.

Zell-Ravenheart, O. 2009. *Green Egg Omelet: An Anthology of Art and Articles from the Legendary Pagan Journal*. Franklin Lakes, NJ: New Page Books.

INDEX

A NOTE ON THE TYPE

THIS BOOK has been composed in Miller, a Scotch Roman typeface designed by Matthew Carter and first released by Font Bureau in 1997. It resembles Monticello, the typeface developed for The Papers of Thomas Jefferson in the 1940s by C. H. Griffith and P. J. Conkwright and reinterpreted in digital form by Carter in 2003.

Pleasant Jefferson ("P. J.") Conkwright (1905–1986) was Typographer at Princeton University Press from 1939 to 1970. He was an acclaimed book designer and AIGA Medalist.

The ornament used throughout this book was designed by Pierre Simon Fournier (1712–1768) and was a favorite of Conkwright's, used in his design of the *Princeton University Library Chronicle.*